Energy Systems in Electrical Engineering

Series Editor

Muhammad H. Rashid, Florida Polytechnic University, Lakeland, USA

More information about this series at http://www.springer.com/series/13509

Vinod Kumar Singh · Akash Kumar Bhoi ·
Anurag Saxena · Ahmed F. Zobaa · Sandeep Biswal
Editors

Renewable Energy and Future Power Systems

Springer

Editors
Vinod Kumar Singh
Department of Electrical Engineering
S. R. Group of Institutions
Jhansi, India

Anurag Saxena
Department of Electrical Engineering
S. R. Group of Institutions
Jhansi, India

Sandeep Biswal
Department of Electrical Engineering
O.P. Jindal Global University
Sonipat, Chhattisgarh, India

Akash Kumar Bhoi
Department of Electrical and Electronics Engineering
Sikkim Manipal University
Rangpo, Sikkim, India

Ahmed F. Zobaa
Department of Electrical and Electronics Engineering
Brunel University London
Uxbridge, Middlesex, UK

ISSN 2199-8582 ISSN 2199-8590 (electronic)
Energy Systems in Electrical Engineering
ISBN 978-981-33-6752-4 ISBN 978-981-33-6753-1 (eBook)
https://doi.org/10.1007/978-981-33-6753-1

This Springer imprint is published by the registered company Springer Nature Singapore Pte Ltd.
The registered company address is: 152 Beach Road, #21-01/04 Gateway East, Singapore 189721, Singapore

Preface

Renewable energy (sustainable power sources) affords the means for a perfect and limitless supply of energy that will become increasingly important to enable us to enjoy a safe and productive existence on planet Earth. They differ from petroleum products by their very nature and abundance, and their potential for use anywhere on the planet. Of greatest importance, however, is that they produce neither greenhouse gases—which cause environmental change—nor contamination. In addition and in contrast to petroleum derivatives, and their related unpredictability, the production costs of renewable energy are falling at a good rate.

As indicated by the International Renewable Energy Agency (IRENA), increasing the use of renewable energy to 36% by 2030 will bring about additional worldwide development of 1.1% by that time (equivalent to US$1.3 trillion), a 3.7% increase in prosperity of 3.7%, and an increase in employment in excess of 24 million jobs, in contrast with 9.2 million today.

This book sums up the key powers driving global change, presents a means by which to assess choices with respect to the degree and movement of progress, and identifies pathways for change. Extraordinary changes in innovation, strategy, financing, and plans of action are globally driving our approaches to energy. Considering these patterns, it is no longer a question of whether power frameworks will be changed; rather, it is a question of how these changes will take place. There are three approaches to dealing with the dynamics of strategy and innovation: versatility, reconstruction, and transformation. Within this framework, we investigate the solutions that have risen as feasible models through which to change how our planet is affected by our need for energy.

Chapter "Renewable Energy and Economic Dispatch Integration Within the Honduras Electricity Market" discusses how power purchase contracts with solar plants produce a strong distortion in generation, especially when compared with an economic firm that follows criteria based mainly on a perfect supply and demand market. There is a cost difference between plants engaged in economic dispatch and companies such as ENEE. In economic dispatch plants are distributed according to the hourly demand and mainly in the order of their costs, taking into account the capacity restrictions of the generators, the transmission lines and other generation systems in which the same criteria of economic generation are followed. ENEE is required to distribute all the output of renewable energy plants, especially those producing photovoltaic (PV) and wind energy. The difference is US$5,143,000/week, which is equivalent to approximately US$23,200,000/month and US$282,250,000/year.

Chapter "Converter/Inverter Topologies for Standalone and Grid-Connected PV Systems" presents a complete outline of the topologies of converters/inverters used in standalone and grid-connected PV systems based on their control procedures. Focusing on converters/inverters used in PV systems, a comparison is made based on the general characteristics of converter topologies. The various converter topologies used to track maximum power are analyzed and presented, and the features of each method of maximum power point tracking (MPPT) are discussed. To further enhance knowledge of these topologies, comparative tables are presented that consider the parameters that will influence the performance of any converter topology, such as losses, efficiency, cost, and so on. The authors also focus on reviewing the most widely used inverter topologies for application in solar PV systems. Numerous inverter topologies are presented thorough an analysis of their various factors: power rating, efficiency, switching time, the number of switches of total harmonic distortion (THD), and the lifespan of the inverters. This should help the reader to choose a suitable converter/inverter topology according to their required applications. This chapter can serve as a "one-stop shop" solution for all researchers and engineers working on converter/inverter topologies, enabling them to work with a particular converter/inverter for their specific application. In addition, the authors expect that this research will prove to be a benchmark for practitioners and researchers in the area of converter/inverter topologies.

Chapter "Mission Profile Oriented Reliability Evaluation of Grid-Connected PV Inverter Considering Panel Degradation and Uncertainties at Indian Location" presents the impact of mission profile, panel degradation, and uncertainties on PV inverter reliability. A case study of a 3-kW grid connected full bridge PV inverter with four 600V/30A insulated-gate bipolar transistor (IGBT) is considered for an evaluation of its reliability. The chapter then presents real-time mission profile data for one year comprising one-minute samples logged at the B V Raju Institute of Technology, Narsapur, Medak, India. A Foster electro thermal model is used to estimate the junction temperature of an IGBT, and a rainflow counting algorithm is used to analyze the junction temperature variations. Using a Monte Carlo simulation, a sample of 10,000 data with 5% variations (uncertainty) is generated. A two-parameter Weibull distribution-based reliability evaluation is then carried out, both at component level and at system level, and PV inverter reliability is analyzed. Results reveal that the mission profile, panel degradation rate, and uncertainties significantly impact the reliability of a PV inverter. The B_{10}, B_{50} lifetimes, with and without the degradation rate, are calculated and compared.

Chapter "Performance Analysis of PI and PR Controller for a Single-Phase PV Grid System with Effective Active and Reactive Power Compensation" discusses applied reactive power management for grid-connected PV networks with different control strategies. The suggested simpler reactive energy controllers will affect impact of the processor measurement such that an inexpensive PV inverter converter can be applied. To show power stability, a limited signal analysis is carried out for the PR system. When evaluating the effects of different active power and reactive power compensations, the most efficient device is a proportional-resonant (PR) controller, which can be used to achieve the optimal active/reactive power regulation over a proposed single-phase grid-connected PV network. The chapter shows that, once compared with the PI system, the feedback from the PR controller can be easily used to enhance the output.

Chapter "Design and Analysis of Fifth-Order Bi-Directional Charger with Vehicle to Grid Application" presents the simulation of a converter design using a PSIM simulator. The various current and voltage waveforms are compared with a bidirectional DC–DC converter without complimentary switching, which shows that the converter design is capable of operating in the soft switching mode, thereby increasing efficiency and leading to a reduction in switching losses. Linear or small-signal models of the converter are required in order to design the digital voltage-mode controller. The chapter continues by discussing the design of the controller, for which the standard pole-placement technique (SPPT) is used, and the Matlab sisotool interactive platform are used, which enables the controller's poles and zero locations to be easily adjusted under stable conditions such as a gain margin > *6 dB*, and a phase margin: 45°–75°.

Chapter "Photovoltaic Inverter Model in Simulink" discusses how the project was instrumental in introducing the team to the iterative process of designing electronic systems for use with power. It was not only a tremendous learning experience, but also exposed the team to the breadth and depth of design considerations, and the problems that are encountered throughout the design and modeling of a grid-tie photovoltaic inverter.

Chapter "Smart Power Management System for Charging Plug-in Hybrid/Electric Vehicles Using Solar PV for Software Technology Park" discusses the role parking garages could play in relation to renewable energy as they afford an ideal space in which solar power capacity may be factored into the electrical matrix. When a driver needs to recharge their vehicle to complete their return journey home, it is advantageous to be able recharge their vehicle while at work. A considerable amount of solar power can be created utilizing sun-oriented boards in parking areas; facilities such as this meet the needs of workers with electric vehicles and a job that requires their attendance during regular business hours. The Nissan Leaf, which uses solar power, has not only been the most cost-effective option to run for more than ten years, but also produces the lowest level of ozone-depleting exhaust products when compared with other options that were researched. Likewise, the impact the Nissan Leaf has on urban air quality is minimal. Since it has a limited travel range, the opportunity to charge the vehicle at work has great worth. There are numerous reasons and advantages related to pushing ahead with the establishment of sun-oriented fueled charging stations, the utilization of which for electric vehicles affords a variety of benefits in terms of natural, social, and monetary perspectives. Sunlight-based charge stations and electric vehicles at present do, and will continue to, profit society in general.

Chapter "Enhancement of IEEE 802.11 Based Network Performance Using Combined Optimization of Parameters for Smart Grid Network" depicts the joint optimization of .... (PHY) and .... (MAC) parameters of IEEE standard 802.11. Diverse set of results are attained for various parameters such as total throughput, MAC delay, and total packet delay. Graphic results illustrate that the throughput and media access delay are higher in case of an .... (RTS) threshold of 16. Tables and figures illustrate that total packet delay is lowest for an adapted EDCA-3 and highest for adapted EDCA-1. Results illustrate that the lowermost value of MAC delay is achieved for adapted an EDCA-3. It is apparent from the results that joint optimization augments the performance of a wireless network. This chapter is anticipated to serve as a directive for joint optimization of various parameters not only for a variety of standards, but also for applications for future research endeavors.

Chapter "Wind Turbines in Energy Conversion System: Types & Techniques" discusses how the provision of renewable energy by means of wind power has seen a swift take-up. Wind turbines employ many different technologies and this chapter evaluates the three main types of wind turbine used since the turn of the millennium. The chapter also discusses how generators require wind-powered installations to respond more swiftly to variations in wind speed, and to have full control over active and reactive powers.

Chapter "Optimal Planning of Reactive Power in Power Transmission System Ensuring System Security Using Probabilistic-CSAJAYA" deals with security achieved through reactive power planning (RPP). Hybridization between two algorithms was rendered to find the solution to RPP. It was seen that combining .... (CSA) with .... (JAYA) could provide satisfactory solutions through which it was possible to establish the optimal set of control parameters within their ranges related to the minimization of total operating costs. The smoothness of convergence curves obtained from a PSODE algorithm implied the performance of the proposed approach with respect to other soft computing techniques. .... (SVC) and .... (TCSC) were successfully installed at weak buses and lines by removing technical and computational burdens. These weak positions were determined through a different analysis, while all constraints were within their limits.

Chapter "An Architectural and Control Overview of DC-Microgrid for Sustainable Remote Electrification" presents an overview of the architectural and control aspects of .... DcMG in the context of remote electrification. There has been a considerable research on various aspects of DcMG. The system is simulated and the results discussed. A comprehensive analysis and comparison of different modes of operation have been presented in the DcMG, which is gaining interest among researchers from a variety of perspectives such as reliability, minimization of power electronics converters and the loss associated with them, and cost effectiveness. There is a great deal of scope for research into DcMG for its value regarding remote electrification in India. The DcMG will play an important role in future energy demand and should be considered as a future grid.

The editors wish to thanks all the authors who have contributed to this book and extend sincere thanks to the Springer Nature Editorial and production team for its constant support. We send our best wishes to our readers!

UP, India — Dr. Vinod Kumar Singh
Sikkim, India — Dr. Akash Kumar Bhoi
UP, India — Dr. Anurag Saxena
London, UK — Dr. Ahmed F. Zobaa
Raigarh, India — Dr. Sandeep Biswal

Contents

Renewable Energy and Economic Dispatch Integration Within the Honduras Electricity Market 1
Pablo Meraz, Wilfredo C. Flores, Harold R. Chamorro, Jacobo Aguillon-Garcia, Alireza Soroudi, Francisco Gonzalez-Longatt, Vijay K. Sood, and Wilmar Martinez

Converter/Inverter Topologies for Standalone and Grid-Connected PV Systems 35
Sudhakar Babu Thanikanti, Dalia Yousri, Dalia Allam, M. B. Etebia, and Karthik Balasubramanian

Mission Profile Oriented Reliability Evaluation of Grid-Connected PV Inverter Considering Panel Degradation and Uncertainties at Indian Location 81
Sainadh Singh Kshatri, Javed Dhillon, and Sachin Mishra

Performance Analysis of PI and PR Controller for a Single-Phase PV Grid System with Effective Active and Reactive Power Compensation 97
Srikanth Sattenapalli and V. Joshi Manohar

Design and Analysis of Fifth-Order Bi-Directional Charger with Vehicle to Grid Application 115
Ashish Kumar Singhal, Narendra S. Beniwal, and Rajesh Kumar

Photovoltaic Inverter Model in Simulink 145
Mostafa Al-Gabalawy

Smart Power Management System for Charging Plug-in Hybrid/Electric Vehicles Using Solar PV for Software Technology Park 161
Kartik Virmani, Y. Raja Sekhar, Akshat H. Mutta, Tarun Sharma, and Naushad Ali

Enhancement of IEEE 802.11 Based Network Performance Using Combined Optimization of Parameters for Smart Grid Network 185
Lipi Chhaya

Wind Turbines in Energy Conversion System: Types & Techniques 199
Bibhu Prasad Ganthia, Subrat Kumar Barik, and Byamakesh Nayak

Optimal Planning of Reactive Power in Power Transmission System Ensuring System Security Using Probabilistic-CSAJAYA 219
Nihar Karmakar and Biplab Bhattacharyya

An Architectural and Control Overview of DC-Microgrid for Sustainable Remote Electrification 241
Yugal Kishor, C. H. Kamesh Rao, R. N. Patel, and Lalit Kumar Sahu

About the Editors

Prof. (Dr.) Vinod Kumar Singh received a Ph.D. in Electronic and Communication Engineering from the Department of Electronics & Communication Engineering, B.U. Rajasthan in 2013, and an MTech in Digital Communication Systems from Bundelkhand Institute of Engineering & Technology, Jhansi, in 2009. He obtained his BTech in Electrical Engineering from the Department of Electrical Engineering, IET Rohilkhand University, Bareilly. He has some 18 years of experience in the field of Electrical and Electronics Engineering.

He is currently Professor and Head of the Electrical Engineering Department, S.R. Group of Institutions, Jhansi UP, India, and is a senior member of the International Association of Computer Science and Information Technology (IACSIT) and the International Association for the Engineers & Computer Scientists (IAENG). He is also a member of the Institute of Electrical and Electronics Engineers (IEEE). He has been working as a coordinator of National Programme on Technology Enhanced Learning (NPTEL), IIT Kanpur, and also as a nodal coordinator of Virtual Labs, Indian Institute of Technology Roorkee.

He is the Vice-Editor in Chief at Blue Eyes Intelligence Engineering & Sciences Publication Pvt. Ltd (BEI-ESP). Prof. Dr, Singh has published more than 200 research papers in international journals published by such as the IEEE, Springer, and Wiley. He is the author of an edited book published by IGI Global (USA) and two reference books. He has tutored and mentored six PhD scholars and more than 30 MTech students. He led the project selected for a financial grant under the Council of Science & Technology Govt. of UP (CST UP). He has delivered expert lectures in many seminars and workshops, and has also organized many interactive workshops and seminars.

He has been appointed as the external examiner for the Ph.D. defence viva in many universities. Prof. Dr. Singh has chaired sessions such as the IEEE Conference (ICACAT-2018) at LNCT Bhopal, the Springer Conference (ICSC-2019) at the Institute of Hydro Power Engineering and Technology Tehri, Uttarakhand, and the International Conference (ICRESE-2016) at Govt. VYTPG, Raipur C.G. He is the reviewer of many well-known SCI journals, and International and national conferences.

Dr. Akash Kumar Bhoi received his B.Tech. in Biomedical Engineering at the Trident Academy of Technology, BPUT University, Odisha, in 2009, and an MTech in Biomedical Instrumentation from Karunya University, Coimbatore, in 2011. He received his Ph.D. from Sikkim Manipal University, India, in 2019. He has been Assistant Professor (Research) in the Department of Electrical and Electronics Engineering at Sikkim Manipal Institute of Technology (SMIT), India, since 2012.

He is a University Ph.D. Course Coordinator for Research & Publication Ethics (RPE), and a member of IEEE, ISEIS, and IAENG, an associate member of IEI, UACEE, and an editorial board member reviewer of Indian and international journals. He is also a regular reviewer of respected journals such as those published by the IEEE, Springer, Elsevier, Taylor & Francis, Inderscience, and so on.

His research areas are Biomedical Technologies, the Internet of Things, Computational Intelligence, Antenna, Renewable Energy. He has had several papers published in national and international journals, and has presented papers at conferences. By 2020, Dr. Bhoi had had more than 90 documents registered in the Scopus database. He has also served on numerous panels engaged in the presentation of international conferences and workshops. He is currently editing several books with Springer Nature, Elsevier, Routledge, and CRC Press. He also serves as a guest editor for special issues of journals published by Springer Nature and Inderscience.

Dr. Anurag Saxena is currently an Assistant Professor in the Department of Electrical Engineering, SR Group of Institutions, Jhansi. His research area is radio frequency energy harvesting (RFEH), improvement in the performance of textile antennas, the effect of radio frequency on the human body, and the specific absorption rate (SAR) calculations of various antenna designs for high frequency ranges such as 4G, 5G, and so on. He has been involved in the field of electrical and electronics engineering for some ten years. He has been a member of the National Programme on Technology Enhanced Learning (NPTEL) and Virtual Labs, SRGI, Jhansi.

Dr. Saxena has published more than 25 research papers in the international journals such as those published by the IEEE, Springer, and so on. He is the author of *Basic Radar Systems*, published by Amazon (Kindle Book Publisher-USA). He has contributed more than ten book chapters to publications by Springer, IGI Global, and so on. He lead a project selected for a financial grant from the Council of Science & Technology Govt. of UP (CSTUP). He has delivered expert lectures in many seminars and workshops, having also organized many interactive workshops and seminars. He is reviewer of many renowned international and national conferences.

Dr. Ahmed F. Zobaa received his D.Sc. from Brunel University London, UK, in 2017. He has spent more than 28 years in academia at the Cairo University (Egypt), the University of Exeter (UK), and Brunel University London (UK). Currently, he is a Reader in electrical and power engineering at Brunel University London. His main areas of expertise include power quality, renewable marine energy, smart grids, energy efficiency, and lighting applications.

Dr. Zobaa is an Executive Editor for the *International Journal of Renewable Energy Technology*, an Executive Editor-in-Chief for *Technology and Economics of Smart Grids and Sustainable Energy*, and an Editor-in-Chief for the *International Journal of Electrical Engineering Education*. He is a registered Chartered Engineer, Chartered Energy Engineer, European Engineer, and International Professional Engineer. He is also a registered member of the Engineering Council, UK; the Egypt Syndicate of Engineers; and the Egyptian Society of Engineers. He is a Senior Fellow of Higher Education Academy, UK; and a Fellow of the Institution of Engineering and Technology, Energy Institute, UK, the Chartered Institution of Building Services Engineers, UK, the Institution of Mechanical Engineers, UK, the Royal Society of Arts, UK, the African Academy of Sciences, and the Chartered Institute of Educational Assessors, UK. He is a senior member of the IEEE.

Dr. Sandeep Biswal received a Ph.D. in Electrical Engineering in the field of power system protection from NIT Raipur, India, in 2019, and an MTech in Power System Engineering from VSSUT (UCE) Burla, Odisha, in 2013. Currently, he is an Assistant Professor in the Department of Electrical Engineering, O P Jindal University, Raigarh, India. He is a reviewer for many well-known journals, among which are the *IEEE Transaction on Power Delivery*, *IEEE Transaction on Power System*, *IEEE Transaction on Circuit and System-I*, *IEEE Transaction on Circuit and System-II*, *IEEE Systems Journal*, *Journal of Electric Power Components and Systems*, *European Transaction on Electrical Power*, *Electric Power System Research*. His research interests include power system relaying and monitoring, microgrid protection, and so on. Dr. Biswal received the POSOCO Power System Award, given by the Indian Institute of Technology Delhi, India, in 2019.

Renewable Energy and Economic Dispatch Integration Within the Honduras Electricity Market

Pablo Meraz, Wilfredo C. Flores, Harold R. Chamorro, Jacobo Aguillon-Garcia, Alireza Soroudi, Francisco Gonzalez-Longatt, Vijay K. Sood, and Wilmar Martinez

1 Introduction

With the increasing integration of non-synchronous generation, the uncertainty of non-dispatchable energy affects the safety and the capability of the grid [1]. Large-scale integration of PV generation may cause an imbalance between the supply and demand of electricity in power systems [2]. Therefore, it is necessary to con-

P. Meraz · W. C. Flores
Universidad Tecnologica Centroamericana, Tegucigalpa, Honduras
e-mail: pablo.meraz@unitec.edu.hn

W. C. Flores
e-mail: wilfredo.flores@unitec.edu.hn

H. R. Chamorro (✉)
KU Leuven, Katholieke Universiteit Leuven, Leuven, Belgium
e-mail: hr.chamo@ieee.org

J. Aguillon-Garcia
Korea Advanced Institute of Scienceand Technology, Daejeon, Korea
e-mail: aguillon@ieee.org

A. Soroudi
University College Dublin, Dublin, Ireland
e-mail: alireza.soroudi@ucd.ie

F. Gonzalez-Longatt
University of South-Eastern Norway, Porsgrunn, Norway
e-mail: fauthor@example.com

V. K. Sood
Institute of Technology, University of Ontario, Ontario, Canada
e-mail: f.gonzalez-longatt@usn.no

W. Martinez
Department of Electrical Engineering (ESAT) at KU Leuven, Diepenbeek, Belgium
e-mail: wilmar.martinez@kuleuven.be

V. K. Singh et al. (eds.), *Renewable Energy and Future Power Systems*, Energy Systems in Electrical Engineering, https://doi.org/10.1007/978-981-33-6753-1_1

sider renewables as a variable in the dynamic economic dispatch. Several authors have addressed this challenge by applying different methods. For instance, the effect of diesel price drop on the economic optimality of these PV-based microgrids is studied in [3]. Moreover, the benefit of varying the dispatch strategy used by the microgrid controller in maintaining the techno-economic optimum configurations of these microgrids as PV-utilizing microgrids is proven.

A model to include demand response in the optimal power flow under wind power uncertainty is formulated as a mixed-integer linear-quadratic problem and evaluated through Monte Carlo simulations in [4].

The impact of the PV plant integration on the Oman transmission systems using an economic dispatch model is investigated in [5]. Simulation results demonstrate that in addition to economic dispatch cost reduction, the PV power plant has positive impacts on system losses.

The authors in [6] conceive a constrained programming to build dynamic economic dispatch model, considering renewable technologies. The thermal limits and spinning reserve constraint are also considered in the model, and the risk is presented directly with probability manner.

The energy supply and demand characteristics with a coordinated control strategy of energy management are proposed in [7]. Additionally, the day-ahead economic dispatch method is applied to the photovoltaic power generation system to ensure the economic, safe , and stable operation of the system.

The interval optimization methodology to solve an uncertain economic dispatch problem for an distribution system operator is proposed in [8].

A grid-connected microgrid consisting of a wind turbine (WT), photovoltaic cells (PV), a battery energy storage system (BESS), a microturbine (MT), a diesel engine (DE), a fuel cell (FC), and several EVs are included in the economic dispatch is studied in [9].

An economic analysis method for hybrid power distribution systems applying an optimal scheduling model for the distribution system in high permeability renewable energy scenarios is presented in [10]. The efficiency at different operational points and converters is calculated based on the measurement data and adopted for loss analysis.

A multi-time-scale robust economic dispatch strategy of a multi-source hybrid power system based on the variable confidence level is proposed in [11]. The deterministic constraints of each time scale are transformed into robust constraints that take the uncertainty into account.

In [12], a hybrid emission economic dispatch model for a solar photovoltaic integrated with thermal generating plants is proposed. A mixed-integer binary programming problem subject to various practical constraints is obtained. A decomposition framework is proposed for solving the problem.

An efficient method based on optimality-condition-decomposition technique is proposed to solve the Dynamic Economic Dispatch problem in real-time environment while considering wind power generation and pool market is presented in [13].

An improved energy management (EM) operation to achieve better energy efficiency in the building is proposed in [14]. In the proposed EM operation, the ED problem is investigated according to several aspects such as the impact of air temperature, thermal resistance (R) of the building envelope, and time horizon (day-ahead ED and five-minute-ahead ED).

A reliability-economic-oriented optimal dispatch model for the distribution network is proposed in [15], which takes into account the prediction range of PV output and the demand.

A mixed-integer algorithm for the optimal dispatch of a storage system, based on the day-ahead PV forecasting is developed in [16]. The optimization objective is the maximization of the total production of the integrated system, according to a requested active power profile.

Hosting capacity (HC) of a distribution system limits the integration of the PVs [17]. To identify the economic efficiency of HC, two boundaries of HC for distribution buses are proposed considering the technology and economy. Then, new evaluation metrics that can identify the HC region for the PV buses are presented. A bilevel optimization dispatch method based on iterative particle swarm optimization for PVHC enhancement is proposed considering the benefits of both distribution network operators and PV owners.

Based on the economic dispatching of a single electric inn and the improved IEEE 33-node case, it is verified that the fragmented energy management mode has a significant impact on improving the optimal operation of distribution network [18]. Numerical results show that the replacement of the battery box for increasing the EV available time has a good effect on peak load shifting of the distribution network and improves consuming of the RES.

Using the forecasted PV power output in economic-load dispatching control (EDC) is essential to maintain the economy and reliability of power system operation. The focus of this paper is placed on an EDC that determines the unit commitment (UC) based on the day-ahead PV generation forecast by numerical simulations [19].

The objective of this study is to make a comparison between an optimal economic dispatch that considers generation and transmission without energy contracts versus the dispatch that includes the 2017 power purchase contracts signed by the country of Honduras.

With the objective of guaranteeing energy supply and reducing dependence on fossil fuels, prior to the reforms of the national electricity sector, the state-issued laws in order to promote generation from renewable sources, granting them a series of incentives among them a price higher than the marginal cost of short-term generation and the obligation on the part of the state to buy all the energy produced.

Power producers soon flooded the sector with several wind and solar plants and power purchase agreements (PPAs) were signed for periods of up to 20 years. Thus, it is estimated that the aforementioned contracts are an obstacle to carrying out an economic dispatch since, by law, these plants must be dispatched despite having higher generation costs.

Section 2 describes the current status of the dispatch of generation plants and illustrates how it is done in the neighboring countries of Guatemala, El Salvador, and Panama. In the same section, some programming techniques that minimize generation costs are described.

Honduras is a country located in Central America with a population of approximately 8.3 million people (INE, 2013) with a GDP of MUSD 20.42 (World Bank, 2015) with an annual growth of 3.6%. The electrical energy sector in Honduras was managed by the municipalities until 1957, when the National Electric Energy Company (ENEE, by its Spanish acronym), was created to take charge of the generation and distribution of electricity in the country. The electricity sector currently has small participation of approximately 11.2% in the national energy balance [20].

Currently, the generation plants dispatch in Honduras is carried out disproportionally. This is mainly due to the fact that the restrictions of the network itself are not taken into account but, on the other hand, the operator (by law) is obliged to dispatch the plants whose prices are considerably higher than other plants.

- Make a comparison between an optimal economic dispatch that considers generation and transmission without energy contracts versus the dispatch that includes the 2017 power purchase contracts signed by the country of Honduras.
- Determine the difference in costs between an optimal dispatch and the one currently performed which considers power purchase contracts.

2 Theoretical Background

Honduras, like other Central American countries, is transforming its electricity market due to new market demands and, above all, due to the Country's policies and plans to define the electric dispatch and generation as a strategic key sector. In order to compare the scope of the Dispatch Centers the characteristics of neighboring countries like Guatemala, El Salvador, and Panama are also discussed.

2.1 Honduran Electrical System

The electrical subsector in Honduras was managed by local municipalities until 1957 when the National Electric Energy Company (ENEE, by its Spanish acronym) was created. Since then it is responsible for the generation, transmission, and dispatch of electrical power in the country.

After the inauguration of the 300MW El Cajon hydroelectric project in 1985, Honduras solved its energy deficit which was mainly due to its dependence on more than 50% of fossil fuels for thermal generation. This hydroelectric complex had an initial cost estimate of 700 million USD and supplied about 69% of the national electricity demand. Honduras then started to export energy to neighboring countries [21].

2.1.1 The Energy Crisis

Due to its annual growth in electricity demand of 8%, Honduras' energy requirements quickly depleted the fairly large available energy supply. Starting in 1992, and with a prolonged period of droughts in the Central American region, poor management of the dam's water reserves, plus technical failures in the reservoir construction which allowed significant water leakage, El Cajon lost much of its water supply.

By 1993 and with the ongoing election campaign, the Honduran administration imposed a minor electrical energy rationing. With the election of a new opposition government in 1994, the new government faced a severe shortage of water supply; hence, it was forced to start an immediate power rationing program with daily outages of up to 12 hours for around eight months (from April to December 1994).

Consequently, 1994 was characterized by a marked power supply crisis with an energy shortage of 120 MW and an acute financial problem, which would have been unmanageable for the ENEE without the full support of the population. In addition, the number of technical faults surged from 15 to 29% due to the financial mismanagement of resources creating a deficit of approximately USD410 million, eliminating the possibility of any growth on generation, transmission, and dispatch sectors, which could have supported the demand in all the consumer sectors.

With the ongoing energy crisis, the government resumed at high cost the operation of the existing obsolete thermal plants which, due to lack of maintenance, had been completely abandoned or had already been partially dismantled and sold. These old plants needed significant repairs and capital injection; this made for a slow restoration process. Under a presidential decree, the government requested from the private sector an urgent investment in thermal plant installations to solve the energy crisis.

The complete absence of a national energy policy crippled the national energetic development. Thus, an electricity subsector framework law was proposed, discussed, and approved in the National Congress; this opened new opportunities for energy generation, transmission, and dispatch to the private sector, de-monopolizing the role of the State in this subsector, and defining medium and long-term priorities for renewable energy sources. Similarly, the creation of the energy cabinet and a council of ministers that outlined the national energy policy was advised by a permanent technical committee. An analysis of the regulatory agenda and the energy crisis timeline is shown in [22].

As seen, the first private generation, transmission, and dispatch participation in Honduras was developed in 1992, when the first private company for generation and dispatch was established: the Roatán Electric Corporation (RECO), to which ENEE sold the Bay Islands. Since then, and between 1985 and 1994, there is an appreciable increase in the share of thermal and purchased energy in the country's electrical energy matrix, and a decrease in the share of hydraulic energy.

In 1994, a law promoting electricity generation based on renewable resources was created, known as the "incentives law". As part of these incentives, the payment of taxes for equipment, materials, and services are exempted. Contracts are signed that require that the ENEE not only must buy the total energy produced, but also pay an extra 10% of the short-term cost for a period between 15 and 20 years [22]. Also, in

January 2014, the National Congress of Honduras approved the new General Law of the Electricity Industry, which fundamentally changes how the ENEE will operate paving the way for private investment in all sectors of the energy market (generation, dispatch, commercialization, and transmission).

2.1.2 Breakdown of the General Law of Electricity Industry

Honduran legislation allowed the creation and operation of an electricity market that recognized the participation of the private sector in generation and dispatch and, at the same time, allowed them free access to the transmission network [22]. By Decree 404–2013, the general law of the electrical industry was created, in which the Energy Regulatory Commission (CREE, by its Spanish acronym), emerged as an entity with functional budgetary independence and appropriate administrative power to ensure its technical and financial capacity.

The four segments which intervene in the new electrical energy framework market are: generation, transmission, dispatch, and commercialization. Before that, those activities were grouped into a single entity: the ENEE. This new energy reform marked the end of monopolies and the freedom of choice for consumers.

The generation part is in charge of generating the electrical energy at their production plants and plugging it into the electrical energy network. These production plants could be of diverse categories: gas plants, conventional thermal systems, nuclear power plants, hydraulic generators, or alternative energy production plants.

The transmission section transports the High-Voltage energy, manages, maintains, and repairs the utility infrastructures. Transmission companies cannot have a participation, either directly or indirectly, in companies that generate, distribute, or commercialize the resulting energy. Transmission companies must give nondiscriminatory treatment to transmission network users. Any installation that is part of the transmission network inside the national territory will be subject to the direction and control of the System Operator.

Dispatch companies manage the Medium and Low Voltage energy, as well as maintaining and repairing the used infrastructure (this entity is responsible for ensuring the quality of electric supply without interruptions). Dispatch companies cannot have generating plants, except in exceptional cases that must be certified by the Electric Power Regulatory Commission (CREE). Dispatch companies that serve isolated systems are exempt from this rule, which may have their own generating plants. These isolated Dispatch companies are part of the National Interconnected System, and are required to set up one or more separate companies to carry out the generation activity. Concerning Dispatch facilities, the municipalities will be responsible for reimbursing the difference between the cost of an aerial and underground work (when buried networks are necessary).

2.2 *Guatemalan Dispatch Center*

In the Regulation of the Wholesale Market Administrator, issued by the Guatemalan government in 1998, Economic Dispatch is defined as "The Dispatch of generation units at low cost in order to guarantee the electrical supply-demand of the National Interconnected System calculated accordingly to the established Coordination Norms". The products and services in the wholesale market are as follows:

- Electric power,
- Electric energy,
- Electric energy transport services, and
- Complementary Services.

This Wholesale Market commercial operations are supported by

(a) An Opportunity Market or Spot Market, for electric energy prospect transactions, with a price established on an hourly basis or defined by the Commission (if it considers necessary to reduce this period rate). In this market, each customer purchases from the group of sellers and the transactions are made at the energy's opportunity fee based on the short-term marginal cost, which derives from the Available Supply Dispatch.

(b) A Term Market, for contracts between Agents or Large Users with terms, quantities, and fees previously arranged. In this market, Wholesale Market agents and large users will freely decide on their contract's conditions. Acquisition contracts of power and electrical energy established before the trade law comes into force will belong to the Term Market. The Term Market contracts must be enclosed within the provisions of the law and its regulations; their commercial and operational coordination will be carried out by the Wholesale Market administrator. These contracts do not have minimum energy purchase amount clauses nor limit the right to sell surpluses.

(c) A Market of Daily and Monthly Power Deviation Transactions. In daily transactions, the differences between the available power and the producing participants' stable power, are settled, valued at the power's reference price according to the estimated monthly amount to be used. In the monthly transactions, the volume differences between the effective stable demand of each distributor, large user or exporter and their firm demand effectively contracted are settled during the corresponding seasonal year.

The methodology for calculating these deviations will be established in the Coordination Standards following the provisions of these regulations. This entity determines the load agenda for the available supply, satisfying the expected demand of the Wholesale Market in a determined time; thus, minimizing the total operative costs taking into account the conditions of a minimum purchase of compulsory energy consumption from existing contracts, transportation restrictions and operational requirements for quality and reliability under the criteria, principles, and methodology established in the Coordination Standards.

The distributor must cover the demand corresponding to consumer participants and offer to allocate that conforming to producer participants. Also, this distributor must consider the presence of interruptible demand and the supply's cost restrictions represented by the Failure Machines.

2.2.1 Methodology of Variable Costs and Availability

The hydroelectric plants will indicate their weekly available power and the expected water supply; for systems with an annual regulation reservoir, the volume of water or the level of the reservoir and the amount of weekly energy available will be announced; thus, the Wholesale Market administrator can compute the water value according to coordination's described methodology; likewise, during the first week of November, they must send the monthly projections (utilized from November to June), of contributions and monthly generation expected for the period to the Wholesale Market administrator and the National Electric Energy Commission.

Generators with thermal systems show their weekly available power, their fuel stock and annually declare the methodology for calculating their variable costs. The weekly importers will indicate in their declaration, the amount of energy and power accessible in connection with the calculating methodology for the corresponding variable costs. Hence, the energy produced by a generating unit will be the result of the economic distributor. The commercial activity of purchasing and vending power and energy in the Term Market will not include conditions that imply restrictions to the economic office. In the case of surplus or shortages, between the consumer and the contracted distributor, will be considered sold or purchased in the Opportunity Market, in the most suitable manner.

2.2.2 Current Contracts

Existing Contracts will be considered as belonging to the Term Market and will be directed following the contractual stipulations contained in alleged contracts, including conditions for a minimum purchase of energy. In any case, everything must be planned within their restrictions, with attention to an Economic Office.

The differential costs from the Existing Contracts, concerning the Power Reference Prices, the Opportunity Market prices of the energy supplied, the power and energy not consumed by the regulated demand of the distributor "and all the available" of the Wholesale Market, will be distributed among the Consumer Participants of mentioned market. The Dispatch of alleged costs will be made proportionally to the consumption of each Consumer Participant. The Wholesale Market Administrator will include these costs in the Monthly Economic Transactions Report. The National Electric Energy Commission will establish through resolution, the necessary mechanism for the implementation of the agreed provisions.

2.2.3 Total Operation Costs

The total cost of the Wholesale Market generation operation is made up of the following quantities:

(a) Variable costs,
(b) Costs for not supplied energy, and
(c) Cost overruns for minimum required energy purchase in the Existing Contracts.

The variable costs associated with the operation of the generating thermal and hydroelectric units refer to the node of the respective system; while those associated with imports refer to the node of the respective interconnection. The Wholesale Market Administrator will calculate the variable cost of each generating unit that is available in the Wholesale Market, by provisions of these Regulations and the Coordination Rules as follows:

(a) For each thermal unit, the costs must be associated with the fuel consumption, the operation costs and maintenance, the machinery start-stops as well as their efficiency.
(b) For each hydroelectric power station with an annual regulation reservoir, the variable cost will be the value of the water calculated by the Wholesale Market Administrator and the cost of operation and maintenance. For the rest of the hydroelectric generating plants, the variable cost will be equal to their respective operation and maintenance costs. The Wholesale Market Administrator will optimize the use of available renewable resources.
(c) For non-hydroelectric renewable generating plants, the variable cost will be at least their respective operation and maintenance costs. The Wholesale Market Administrator will optimize the use of available renewable resources as well.
(d) For each amount of imported electricity, the variable cost will be the value calculated according to the methodology reported by the importer according to the generation technologies described in the previous paragraphs.

2.3 *El Salvador Dispatch Center*

According to the operating regulations for the transmission system and the Wholesale Market based on production costs (Official Gazette of the Republic of El Salvador, 2011), economic Dispatch is defined as: "Generated units log and/or the generating group to be logged", which results from minimizing operating costs and deficit for a given demand to be supplied.

The purpose of the operation scheduling is to determine, by the Transmission Unit, the Dispatch plan of the generating structures, whose operation in coordination with the operation of the transmission systems results in a minimizing cost operation and deficit of the electrical system, subject to the fulfillment of the requirements of service's quality and security.

Three agenda types will be carried out according to the horizon covered by the indicated plan, namely an annual schedule, a weekly schedule, and a daily schedule, also called pre-Dispatch.

2.3.1 Annual Timetable

The annual calendar must cover fifty-two weeks from the first day of application and will be carried out accordingly to a weekly detail. This schedule is updated monthly.

2.3.2 Weekly Timetable

This must cover seven days from the first day of application and will be carried out according to a scheduled detail.

2.3.3 Daily Timetable

The daily schedule (also called pre-Dispatch), is drawn up daily, carried out accordingly to a schedule plan, and covering a twenty-four-hour period corresponding to the day after which it is carried out. This schedule must be updated under the provisions of the Regulations.

2.3.4 Real-Time Operation

The actions carried out by the Transaction Unit to comply with the plan established in the daily schedule through the operation and coordination instructions given by it to the participants, in compliance with the quality and service safety requirements, will be called operation in real time.

2.3.5 Objective of Annual, Weekly, and Daily Programming

The transactions units will schedule the operation with the objective of determining the systems dispatch and the operation of the transmission facilities with the intention of minimizing the total costs of operation and deficit in the electrical system, preserving the security and quality of supply. This operation will be carried out independently of generation and transmission facilities ownership, and independently of the commercial commitments of the market participants.

2.4 Panamanian Dispatch Center

According to Panama's National Dispatch Center (formerly known as CND, by its Spanish acronym; currently defined as ETESA), "The Electric Market is the area where short, medium- and long-term commercial transactions are carried out between participants, for the sale and purchase of energy or power. In other words, it is a market that, like others, works by balancing supply and demand. Therefore, the electric market is based on competition between generating companies, and aims to increase the quality of supply, the environment and self-regulated free market prices".

The Panamanian state has a set of rules for the Wholesale Electric Market, (the Republic of Panama, 2002). The products that are traded are Energy and Power, and the services provided are transmission, auxiliary facilities, and operating stations.

The Contract Market is defined as the area where medium or long-term commercial transactions are carried out between Participants, for the purchase/sale of energy and/or power with terms, quantities, conditions, and prices resulting from agreements between parties.

Consumer Participants' purchases are guaranteed of supply, which is achieved through the Contract Market. Each Distributor must guarantee supply by Purchases in the Contract Market; their obligations in the contract are defined in the Commercial and the Purchase Rules.

The Occasional Market is the area where short-term hourly energy commercial transactions are carried out, which allow to satisfy the surpluses and shortages that arise as a result of the departures between contractual commitments and the reality of consumption and generation. Every hour, the CND must manage the energy transactions of each Consumer Participant. The energy transactions and the calculation of price in the Occasional Market will be carried out hourly. The National Public Services Authority may reduce this calculation phase in the extent that the measurement's system allows it, and the commercial and operational reality shows the need for a smaller calculation step.

2.5 Economic Dispatch Model

The Economic Dispatch in an electrical power system consists of determining the level of active power generation of each available plant so that the generation costs, in the medium term, are minimal. To achieve this, it is necessary to take into account certain restrictions imposed by the system: i.e., the need for the sum of the powers of all the plants to be equal to the power demand (including losses), the capacity restrictions of each unit, the limitations in the transmission lines, and the quality of service and security [23].

In mathematical terms, the problem can be modeled very precisely, i.e., the objective function represents the total cost of supplying a given load. The aim of the solution is to minimize this function respecting the constraint that generation must be equal to demand [24].

2.6 *Optimization Methods*

The purpose of this present work is focused on the realization of an economic office, applying some optimization techniques. These techniques, by order of complexity, are discussed next.

2.6.1 Linear Programming

Linear programming is a very powerful tool, its objective is to establish the values of certain variables that cause some variable to be maximized or minimized; for example, maximize profits, production, benefits, or the acceptance of a product or minimize costs, defective products, risks, or the time required to produce a product. These well-applied maximizations/ minimizations can save a reasonable amount of money for the energy establishments (Hillier, 2015). Linear programming uses a mathematical model to describe the problem, where the linear adjective means that all the mathematical functions of the model must be linear functions.

In this case, the word programming does not refer here to computational terms. In essence it is synonymous with planning. Therefore, linear programming involves planning activities to obtain an optimal result, i.e., the result that best achieves the specified goal according to the mathematical model, among all feasible alternatives.

Furthermore, a very efficient solution procedure called the simplex method is available to solve these linear problems, even large ones. These are some reasons for the tremendous effect of linear programming in recent decades. The standard model is given by

$$Z = \sum_{i=1}^{n} c_i x_i \tag{1}$$

and is bounded by the following limits:

$$\begin{gathered} \sum_{i=1}^{n} a_{j,i} x_i \le b_j \\ x_j \ge 0 \end{gathered} \tag{2}$$

In the previous equation, x_i is the variable for which we want to find a value such that Z is a maximum; c_i, a_j, i, and b_j are constants where their values are determined

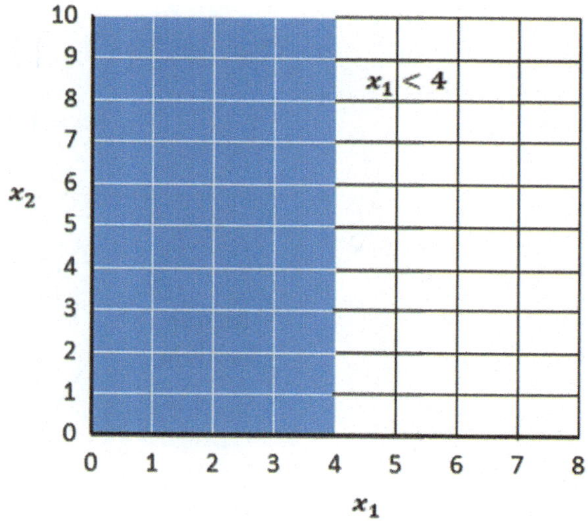

Fig. 1 Optimization by graphical method. The shaded area represents the possible solutions, once the restrictions have been applied that both x_1 and x_2 must be positive and also, x_1 is less than or equal to four

by the characteristics of the problem to be solved, and m is the amount of constraints. Hence, in this method, all variables must be positive.

For the optimal solution of the proposed problem, it is very illuminating use a graphical method to find the answer. For example, if the system needed to find the solution [25] of the following:

$$\begin{aligned} Z &= 3x_1 + 5x_2 \\ 0 &\le x_1 \le 4 \\ 0 &\le x_2 \le 6 \\ 3x_1 + 2x_2 &\le 18 \end{aligned} \tag{3}$$

It is observed that this problem only depends on two variables; therefore, it is reasonable to choose the graphical method (if there were three variables, the graphs will start to be complex three-dimensional graphs). In the graph, each variable is placed on one of the two axes. The constraints are the shaded area that precisely limits the number of possible solutions. Initially, the solutions are on the Cartesian plane, since both x_1 and x_2 are positive the solution is limited to the first quadrant.

Figure 1 illustrates that the shaded area depicts the possible solutions after considering that both x_1 and x_2 are positive and that x_1 is also less than or equal to four.

In Figure 2, the rest of the restrictions have been applied and, by consequence, the shaded area has been reduced. Each point in the shaded area meets the constraints; consequently, it is just needed to find which point maximizes the given function.

Now, we must find the maximum value of the expression shown in Eq. 2.4, for this, we can assign Z an arbitrary value and find the corresponding values of x_1 and x_2, these obviously form a straight line that is illustrated in Figure 3, as there are

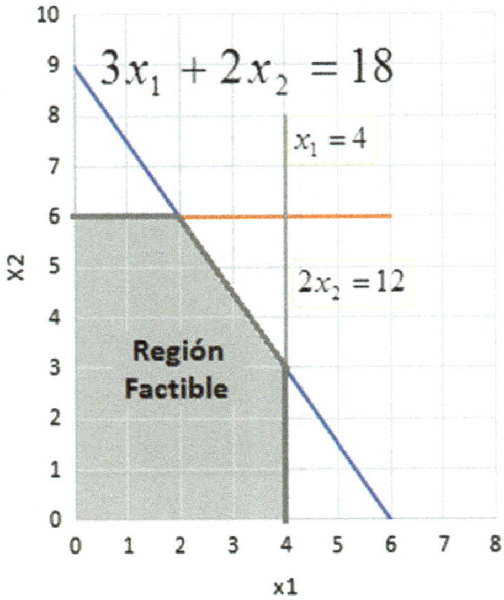

Fig. 2 Constraint Graph. The shaded area represents possible solutions once all constraints have been applied

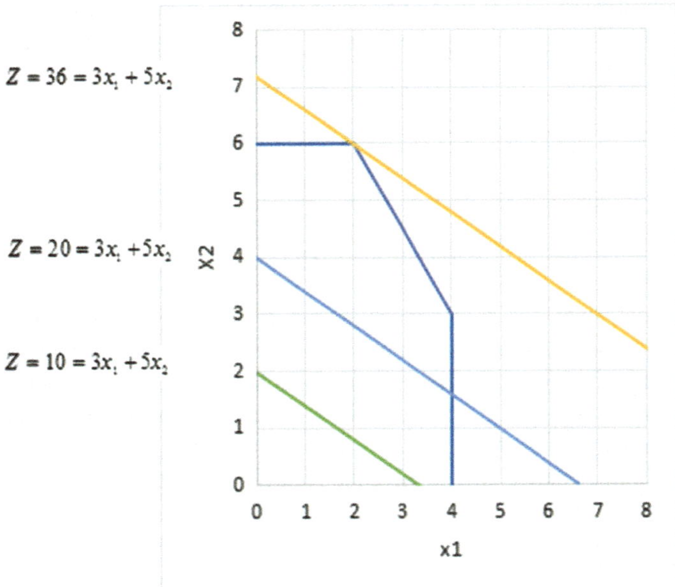

Fig. 3 Maximization lines corresponding to $Z = 10$, 20, and 36 in which all of them are parallel. The solution occurs for $Z = 36$, $x1 = 2$ and $x2 = 6$. An additional line for a value of Z greater than 36 would depart from the shaded area and although it would maximize Z, it would not meet the restrictions

many points that belong to both the straight line and the shaded area, it is possible to try a higher Z value (for example, 20), which also produces a straight line which is parallel to the first one and, thus you can continue increasing the Z value drawing parallel lines being careful not to leave the shaded area.

Following this process, it is found that $Z = 36$ is the highest value that can be reached.

2.6.2 Mixed-Integer Programming

A mixed-integer linear programming problem (MIP), is a linear problem (LP), with some integer variables. These types of problems are more difficult to solve than linear programming problems. The first solving algorithm was formulated by Ralph Gomory in 1958 [26].

There are several solution methods, among them are linear relaxation and discretization, exhaustive enumeration, branch and bound, cut planes method, branch and cut between others [27].

2.6.3 Decomposition Techniques

Benders' decomposition algorithm, proposed in 1962, has been applied very successfully to a range of problems whose solution is extremely complicated [28]. The algorithm was proposed with the main objective of addressing highly complicated problems which, once the variables have been rearranged, lead to problems that are significantly much easier to solve. This method is also known as variable decomposition and external linearization. Currently, it is one of the most efficient and exact methods that reduces considerably the computational effort. Its applications can be found in various fields including planning, transportation, telecommunications, energy, resource management, and chemical processes.

The problem previously stated can be solved using the Benders' decomposition as follows:

$$f^T y + c^T x \tag{4}$$

bounded by

$$\begin{aligned} Ay &= b \\ By + Dx &= d \\ x &\geq 0 \\ y &\geq 0 \end{aligned} \tag{5}$$

for integer variable values.

2.6.4 Nonlinear Programming

Fundamentally, nonlinear programming is very similar to linear programming in terms of formulating a problem, the difference is that in the latter case at least one of the constraints or the minimization function itself does not show a linear behavior [29]. Other differences with linear programming are that the solution is not always located at an extreme imposed by a restriction. It is feasible to find local optimums which might contain non-convex zones [30]. For nonlinear programming problems occasionally, the exact solution is not found, but through an iterative method it is possible to converge toward it. The process ends when a solution is found that, under certain practical criteria, is sufficiently close to the solution [31].

3 Methodology

This section describes the methodology to be used to achieve the objectives set out in Sect. 1. In summary, this methodology consists of collecting data related to the optimization process which must be adapted to introduce them to the calculation in the form of data, equations, and constraints. Afterward, performing the necessary calculations on the General Algebraic Modeling System optimization program (GAMS) [32], and finally, the data will be modified to make additional executions to meet each of the objectives of this thesis. In the following sections, each of the processes of the aforementioned methodology is described in more detail. Figure 4 illustrates the processes required according to the objectives of this study.

3.1 Data Acquisition

To fulfill the required objectives of the study, the correct data must be acquired as follows:

- National demand data: Typical daily demand curve for business days, holidays, Saturdays, Sundays. Monthly and annual growth rate, seasonality of demand.
- Generating plants,
- Technology,
- Location,
- Nodal energy price,
- Generation cost curve,
- Minimum/maximum power,
- Maximum energy for one period,
- Starting characteristics: Cost, minimum operational hours, minimum idle hours
- Transmission lines: Voltage level, line impedance, maximum capacity, and
- Generation Incentives Law (Decree 138–2013).

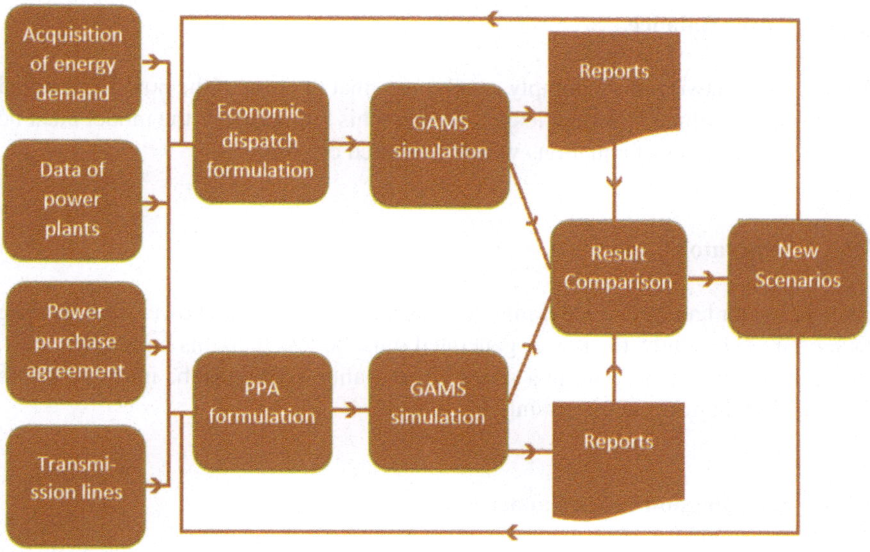

Fig. 4 General Outline of the Investigation

3.2 *Equation Formulation with Economic Dispatch*

3.2.1 Cost Equation

The objective is to minimize the costs of generating electricity. This involves writing a formula that represents the generation costs. By posing a nonlinear programming problem, it is possible to find the values of the variables that make the expression minimal. The following formula is the sum of the generation costs in each of the hours of the period, being the cost of a particular hour, in turn, the sum of the generation costs of each generator corresponding to the generation level according to the dispatch. The above can be written mathematically as follows:

$$cost = \sum_{ih} \sum_{igen} C\left(g_{igen}, ih\right) \tag{6}$$

In which i_h is an index for each of the hours of the period log, i_{gen} is the index that executes through each of the generators, gigen, ih is the optimal energy Dispatch according to the generator's model for one hour in particular.

3.2.2 Power Balance

An electrical network must comply, at all times, that the sum of the power generated is equal to the demand plus the losses. To meet this requirement, the model must be limited with a series of equations which are called constraints.

3.2.3 Generator Capacity

Each generator has specific operating limits such as a maximum power limit. During each hour of the study, the power generated must be less than this mentioned level. All of the above limits result in a series of inequalities that must be introduced into the model in the form of constraints.

3.2.4 Transmission Lines Capacity

Likewise, the transmission lines have both power and voltage drop limits. These also result in more constraints for the model.

3.2.5 GAMS Calculation Tool

The selected tool to solve the mathematical optimization problem is the General Algebraic Modeling System (GAMS), designed to model and solve mixed linear, nonlinear, and integer optimization problems. The system is designed for complex large-scale modeling applications and allows the user to build large models that can be adapted to new situations. The system is available for use on different computing platforms.

Models are exchangeable from one platform to another. GAMS was the first algebraic modeling language and its structure is similar to that used in other fourth-generation programming languages. GAMS contains an integrated development environment.

3.2.6 Data for GAMS Model

The data must be entered in a spreadsheet style due to practical management. Later using a computer program written in Python, these will be converted to suit the GAMS program. The use of GAMS program to solve a nonlinear programming problem is illustrated below:

$$F = 13x_1 + 6x_1x_2 + 5x_2 + 1/x_2 \tag{7}$$

subject to

$$\begin{aligned} 2x_1^2 + 4x_2 &\leq 90 \\ x_1 + 2x_1^3 &\leq 75 \\ 8x_1 - 2x_2 &\leq 61 \\ 8x_1, x_2 &\geq 0 \end{aligned} \tag{8}$$

3.2.7 GAMS Program Execution with Economic Dispatch and Contracts

It is necessary to execute two models: (a) This simulates a pure economic dispatch and (b) This considers power purchase contracts (PPA), which consist of contracts between two entities; one entity which generates electricity and the other entity which needs to buy its electricity.

Through the PPA, all commercial terms related to energy sales are defined, including the definition of the initial date of commercial operation, energy delivery plan, penalties for non-compliance, payment terms, and termination. In economic dispatch, the order of merit of generating plants will be governed by their real generation costs. While considering purchase contracts, the priority will be given precisely by the contractual clauses imposed in such distributing plants.

3.2.8 Comparative Results

With the results obtained from GAMS mentioned in Sect. 3.2.6, a comparison will be made between an economic office vs. the generation that considers contracts, which is the objective of this study. From these results, other important conclusions will also be drawn.

3.2.9 Sensitivity Analysis

To meet the specific objectives of this study, it will be necessary to modify the data input to create several scenarios which include demand variations, expiration of contracts with plants, change of technology, etc.

3.2.10 Variable Identification

The variables can be of two types, i.e., dependent or independent.

Dependent variable:

- Generation Cost.

Independent variables:

- n – Generator quantities

- G_i – Maximum i-generator power,
- a_i, b_i, c_i – i-generator's square cost curve,
- n_t – Transmission lines,
- T_i – Maximum i-transmission line's capacity,
- Demand hour, and
- D_h – Demanded energy per h hour.

4 Analysis and Results

4.1 Economic Dispatch

As mentioned in previous sections, economic dispatch consists of executing a generation plan that in principle satisfies the instantaneous demand at the lowest possible cost. The period considered for this study is hourly during one week to cover different scenarios, both of generation availability and of variations in demand. This includes a week that corresponds to a month during the dry season, and another during the rainy season. The months considered are March and August respectively.

The dispatch considered is DC Dispatch, which has the advantage of being a method that follows a linear and non-iterative power flow algorithm. A feature of this method is to assume that the magnitude of the voltage is measured as 1 p.u. DC power flow is calculated as the active power flow in transmission lines and transformers. This method does not allow the voltage drop calculation at the nodes, nor of the reactive power flow. The advantage of this method lies precisely in that its formulation and solution is much simpler and faster.

It is important to consider the technical restrictions of the system, one of them being the limitations in the transmission lines' capacity, which will be described in the next subsections.

4.2 Per Unit System

For the subsequent analysis, the base constraints have been selected as follows: $Sbase = 100\text{MVA}$, $Vbase = 230\text{kV}$, $Zbase = 529\Omega$ where $Sbase$ is the base apparent power, Vbase is the base voltage level, and $Zbase$ is the base impedance.

4.3 Transmission Lines' Capacity Restrictions

Figure 5 shows a simplified transmission line, which is connected to two nodes of the network, that is, two substations identified node i and node j.

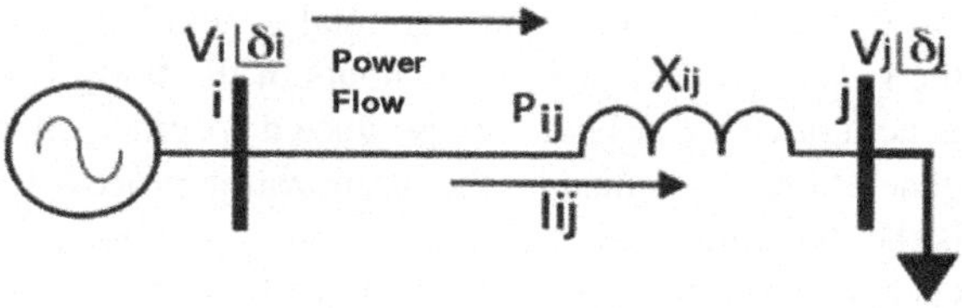

Fig. 5 Section of an electrical system showing the relationship between the nodal voltages and the Power flow

The current between node i and node j is given by the following equation:

$$i_{ij} = \frac{V_i\angle\delta_i - V_j\angle\delta_j}{X_j} \tag{9}$$

In the above equation, i_{ij} is the current flowing from node i to node j, $V_i\angle\delta_i$ is the voltage of node i or j depending on the index and it specifies both its magnitude and its phase angle, and X_j is the reactance of the line. The power flowing from node i to j is

$$S_{i,j} = V_i\angle\delta_i I_{i,j} \tag{10}$$

where, $S_{i,j}$ is the power flow between buses i and j, $V_i\angle\delta_i$ is the voltage angle and magnitude in the bus i, and $I_{i,j}$

As seen in Eq. 10, the sine function disappears since this function is equal to the argument, when the argument is very small.

4.4 Optimization Problem Statement

The optimization problem can then be summarized as follows:

$$\sum_h \sum_i C_{i,h} \tag{11}$$

bounded by:

$$0 \le g_{i,h} \le P_{max,i} \tag{12}$$

$$\sum_h g_{i,h} \le Energy \tag{13}$$

$$g_{n,h} = f_{n,h} + d_{n,h} \tag{14}$$

$$\left|p_{i,j,h}\right| \le P_{max,L} \tag{15}$$

Equation (11) represents the total generation cost, which is the sum of each of the generators and each of the hours of the study. Specifically, C_i, h, represent the generator cost i at hour h to produce a specific generation. In Eq. (12), g_i, h is the generation determined by the optimization process for generator i at time h. $P_{max,i}$ is the maximum power capacity of the generator i; Eq. (13) specifies that the total energy produced by generator i must be less than or equal to a certain amount of energy (Energy) according to the generation plan; Eq. (14) is interpreted as follows: for each 58th. nodes considered in this study and each of the 168 h that comprise the measured period (one week), it must be satisfied that the energy injected into the node due to the presence of one or more generators (if there is not, then it is zero), must be equal to the electric demand at the node plus the sum of all the powers that flow from that node to other nodes (if energy flows goes into it, then it has a negative sign). In this equation:

- $g_{n,h}$ is the generation injected to node n at hour h. It is possible that a node does not have any generator connected (in this case g_n, h would be zero) or have 1 or more.
- $f_{n,h}$ is the net flow of energy from node i to other nodes. If it leaves the node it is considered positive, if it enters the node it is considered negative.
- $d_{n,h}$ is the demand of node n, if a particular node does not have Dispatch lines, then its demand is zero.

Equation (15) states that each transmission line has an energy amount limit that can flow through it, where P_i, j, h represents the power flow from node i to node j at hour h; this flow can be positive or negative.

4.5 Data for GAMS Program

4.5.1 Indexing

To dimension the parameter tables, it is necessary to create indexes. The indexes defined in this application are shown below:
where i_{gen} is an index related to generators, i_h is the hour index since this study considers a week as the study time, i_h takes values between 1 thru168 which is precisely the number of hours in a week; i_{coef} refers to the coefficients of a quadratic polynomial of costs of each generator, i_{coef} is the index related to the coefficients of the polynomial of generation costs, i_{lines} is related to the transmission lines, i_{nodes} to the number of nodes and i_{th} is related to the voltage angles (θ), at each of the nodes. In the present study, the eleven most important generators of the electrical system are considered, the notation "/1*11/" for i_{gen}, indicates precisely that this index will take values between 1 thru11. Table 1 shows the data for these plants:

For the cases where the generator has a constant cost per MW (independent of the generated power), then only the coefficient of degree 1 is used for the polynomial,

Table 1 Main plants in the network, 2017

Item	Name	Technology	Installed Power (MW)
1	EL CAJON	Hydroelectric	300
2	CANAVERAL RIO LINDO	Hydroelectric	109
3	ENERSA	Internal combustion	230
4	LUFUSSA	Internal combustion	323.5
5	ELCOSA	Internal combustion	80
6	BECOSA	Carbon-based	60
7	HONDURAS HPGC	Biomass	45
8	MESOAMÉRICA	Eolic	125
9	SAN MARCOS DE COLÓN	Eolic	50
10	VARIOS SOLARES EN EL SUR	Solar	319
11	EMCE	Internal combustion	60

Table 2 Coefficients of generation costs. 2017

Item	Name	a (US/MW2)	b (US/MW)	c (US)
1	EL CAJÓN	0	0	0
2	CAÓAVERAL RIO LINDO	0	0	0
3	ENERSA	1.1927E-2	69.47	56.15
4	LUFUSSA	6.370E-2	56.44	472.08
5	ELCOSA	0	86.052	0
6	BECOSA	0	29.88	0
7	HONDURAS HPGC	0	50	0
8	MESOAMóRICA	0	0	0
9	SAN MARCOS DE COLÓN	0	0	0
10	VARIOS SOLARES EN EL SUR	0	0	0
11	EMCE	0	102.269	0

with the rest of the coefficients taking the value of zero. Table 2 shows the mentioned coefficients.

Table 3 Cost coefficients in p.u.

Item	Name	a (USD)	b (USD)	c (USD)
1	EL CAJÓN	0	0	0
2	CAÑAVERAL RIO LINDO	0	0	0
3	ENERSA	119.27	6947.97	56.15
4	LUFUSSA	637.08	5644.80	472.08
5	ELCOSA	0	8605.20	0
6	BECOSA	0	2988	0
7	HONDURAS HPGC	0	5000	0
8	MESOAMóRICA	0	0	0
9	SAN MARCOS DE COLÓN	0	0	0
10	VARIOS SOLARES EN EL SUR	0	0	0
11	EMCE	0	10226.9	0

4.5.2 Polynomial of Generation Costs in P.U.

On the p.u. system for the coefficients of the polynomial cost there are changes on the coefficients as a is multiplied by $Sbase2$, b by $Sbase1$ and c by $Sbase0$, being as follows (Table 3):

4.5.3 Power Demand

Next, the demand data corresponding to the week of study is introduced, in which two weeks have been considered corresponding to a dry when there is no rain (March 2017), and to a humid month where there are very frequent rains (August 2017). The demand consists of 168 values corresponding to the number of hours a week has. With the aim of obtaining the data for these weeks, demand patterns from previous years, system load factors, and projections of demand growth of 5% per year were taken into account.

To differentiate the demand between business days, variations were generated for each day following a normal Dispatch curve. Figure 6 shows the demand curves for March 2017.

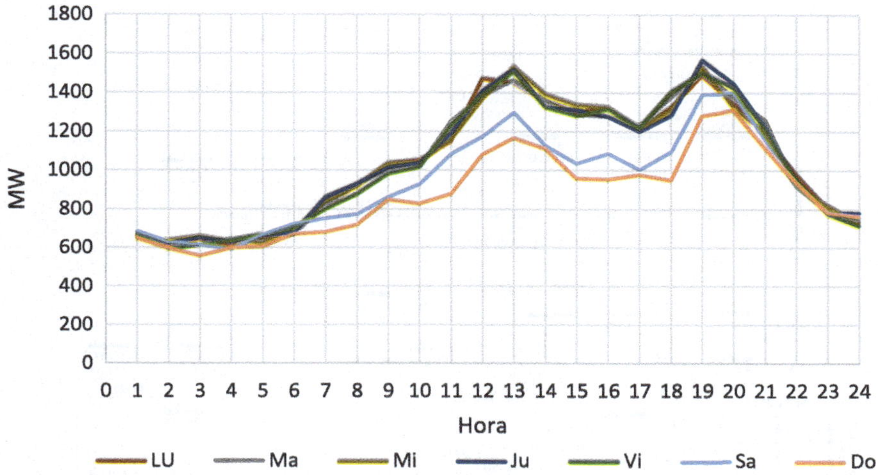

Fig. 6 Demand Curves of the National System Corresponding to a Week in March 2017

4.5.4 Plant Generation Capacity

The GAMS program must also be presented the data corresponding to the plants' capacity, as illustrated below:

Once again, it is noticed that this information is entered into the customary unit system.

4.5.5 Restrictions on Produced Energy

Likewise, it is necessary to introduce the plants' energy limits; some plants have a high generating factor, close to 100%, and the only limitation is their capacity multiplied by 168 h; but others such as hydroelectric plants for reservoir operation reasons and hydrological contributions have an energy limitation such as photovoltaic and wind power have their limitation due to its intrinsic relatively low generating factor.

4.5.6 Transmission Lines

On the network system, 78 transmission lines corresponding to three of the highest voltage levels 230, 138, and 69 kV have been identified. The following figure shows their network dispatch (Fig. 7).

Each of these transmission lines has a reactance associated with it, as mentioned earlier; this parameter is related to the maximum line capacity. The following illustrates how this parameter is presented into the GAMS program:

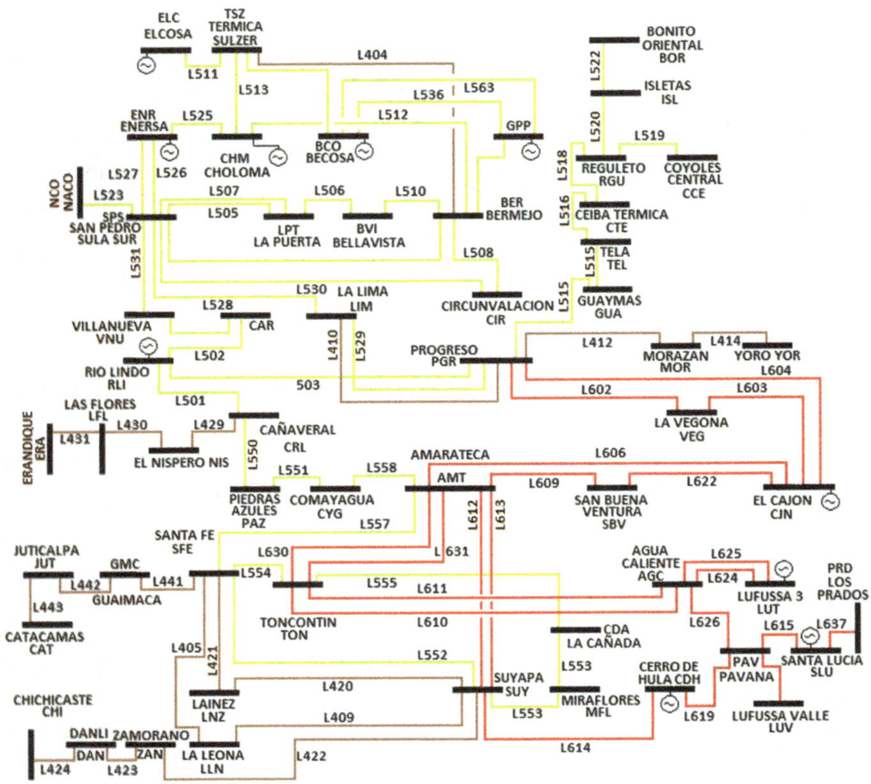

Fig. 7 National electrical system with voltage levels of 230kV, 138kV, and 69 kV in Red, Yellow, and Brown, respectively

To determine the demand at each node, the demand factor parameter has been defined based on the number of dispatch lines per node.

4.5.7 Objective Function in GAMS

The objective function is introduced in GAMS as follows:

$$cost = \sum_{i,h} \left(\sum_{igen} \left(a_{igen} g_{i,h}^2 + b_{igen} g_{igen,ih} + c_{igen} \right) \right) + TDk \qquad (16)$$

where a_{igen}, b_{igen} and $cigen$ represent the quadratic polynomial of costs g_{igen}, i_h is the power generated by the i_{gen} generator in hour i_h.

Total Deficit (TD): This represents the sum of a probable deficit used as a last resort to satisfy the power balance at each node. The model avoids using deficits since it has a higher cost than the most expensive generator in the system by the value of USD153.40/MWh (in the model it is 15340 since the power is p.u.), as is the cost of energy imported from neighboring systems.

This formula corresponds to Eq. (16) expressing that the cost is equal to the sum for each of the 168 h of the product of the sum of generations and the cost of generation. Additionally, it increases with the cost of the deficit at the import price.

In GAMS, the sum is done through the sum statement, the expression "=E=" stands for equal; after the optimization process in which GAMS has determined the generation value for each generator in each hour, it is possible that the same software may determine the marginal hourly-cost value through the following equation:

The marginal hourly-cost must be calculated precisely for each hour. The previous equation performs this calculation using the smax function, which determines the maximum value of a series of indexed values, in this case the indexed values are the unit costs (USD/MWh), for each of the generators at a particular time. As it is possible to see, there are two problems: the first is that the cost function does not produce unit costs but rather global costs; and secondly, the generators that were not distributed should not be considered to determine the marginal cost. To obtain the unit cost, the global cost is simply divided by the generation (producing the potential error of dividing by zero), that is why in the formula, a very small value of 0.001 is added to the generation. The same formula guarantees that generators that do not produce energy do not participate in the calculation of marginal cost. This is achieved through the sign function, which in the case of zero generation, returns values less than or equal to zero. The marginal cost for each hour is calculated taking into account the cost of the plant with the highest cost that was distributed at that hour. Power systems then, tend to distribute the generators with lower costs first and then distribute the generators with increasing costs. The overall marginal cost can be easily calculated once the marginal hourly-costs are available.

4.5.8 Capacity Restrictions

Regarding the capacity constraints represented in Eq. (12) originally only 11 equations were written, each one expressing that, among the 168 generation values the maximum value of each generator must be less than or equal to its maximum power. According to carried out tests, it was found that in addition to an increasing processing time, the results were not as expected. Therefore, it is was preferred to introduce several individual restrictions. Since there are 11 generators and 168 h, this produces 1, 848 restrictions. Due to a large number of inequalities, these were generated automatically using a spreadsheet program suite.

4.5.9 Energy Balance Restrictions

Next, the energy balance constraints were introduced at each node which has been synthesized in Eq. (14) Taking into account that there are 58 nodes and a balance must

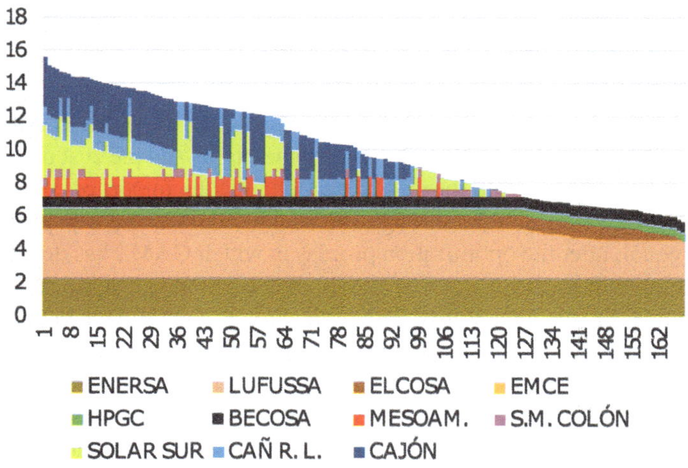

Fig. 8 Demand Curves of the National System Corresponding to a Week in March 2017

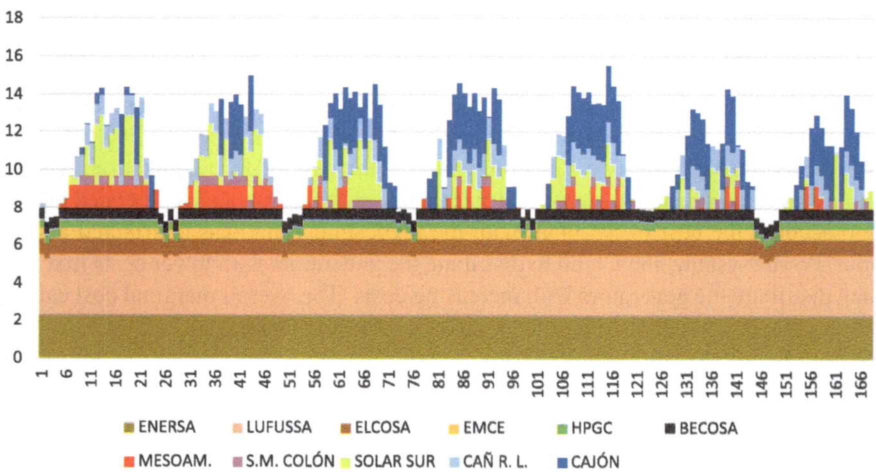

Fig. 9 Demand Curves of the National System Corresponding to a Week in March 2017

be satisfied for each of the 168 h then, there are $58 \times 168 = 9744$. These equations are not identical since each node is connected to different nodes. To write this large number of restrictions, a Python language program was created (Figs. 8, 9 and 10).

4.5.10 Capacity Restrictions on Transmission Lines

Each transmission line has an associated maximum power capacity that can circulate through it. In Sect. 4.2, it was established that the power between two nodes is proportional to the difference between the angles of the voltages and inversely

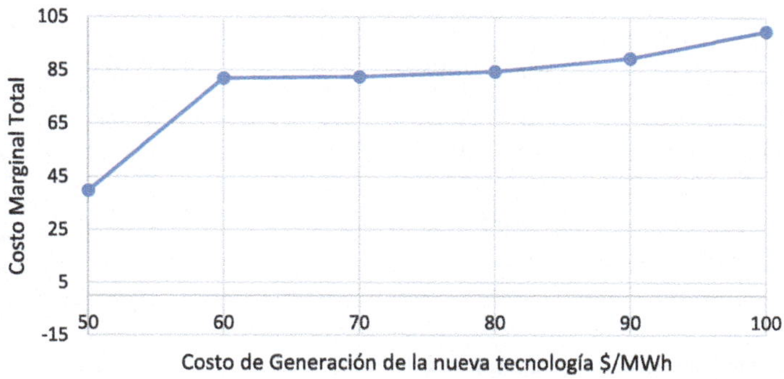

Fig. 10 Demand Curves of the National System Corresponding to a Week in March 2017

proportional to the reactance between them. The corresponding constraint is then stated as follows:

$$\frac{\delta_i - \delta_j}{X_{i,j}} \leq P_{maxL} \tag{17}$$

where $PmaxL$ is the transmission line capacity. The above restriction must be satisfied for the 78 lines identified in this study and in each of the week's 168 h. It is also considered that the flow can be positive or negative, so the number of restrictions are as 78 × 168x2 = 26,208 inequalities. Aided with a spreadsheet suite, the aforementioned restrictions were generated, below is a small portion of these equations.

4.6 *Simulation Results*

4.6.1 Economic Cost Dispatch According to GAMS Program

The most important results are as follows:

- Generation Cost: USD8, 266, 600.00,
- Dispatch cost at marginal cost USD14,521,000.00,
- Energy distributed: USD168,605 GWh/week, and
- Average energy price: USD 86.12/MWh = USD0.0861/kWh.

Summary of energy generated by the plant:

The following two figures show how demand is distributed on a classified demand curve and a normal dispatch curve.

The maximum p.u. demand being approximately 1580MW; ENERSA has been embedded in the base since, as seen, it maintains a constant generation during the study period. Due to their low relative prices, BECOSA and HPGC also maintain their generation throughout the stated period. The plants with energy limitations appear at peak hours. EMCE, due to its high costs, was not distributed in this figure.

Table 4 Comparison between generation in a dry and a wet month

	March, 2017	Agosto, 2017
Generation Cost US$	8266600	10218000
Dispatch cost at marginal cost US$	14521000	19167000
Dispatch Energy GWh	168.604	184.5329
Average cost $/MWh	86.12	103.87

Table 5 Comparison between generation in a dry and a wet month

Central	March	August
EL CAJÓN	18459.8100	18458.3
CAÑAVERAL RIO LINDO	10789.9	10789.7
ENERSA	38640	38640
LUFUSSA	47942.5	54202.400
ELCOSA	12810.7	13440
BECOSA	10080	10080
HONDURAS HPGC	7560	7560
MESOAMóRICA	6700.02	6699.6
SAN MARCOS DE COLÓN	2852.8	2850
VARIOS SOLARES EN EL SUR	12770.01	12769.8
EMCE		9040.7000
Total	168605.74	184530.5

4.6.2 Dispatch Corresponding to Rainy Month August 2017

The following Tables 4 and 5 shows a comparison between the most important results when the computed model is executed with a month considered dry and another considered as rainy month.

From Table 4, we conclude that the average generation cost remains approximately constant despite an increase of approximately 10.8% in energy. The following Table 5 compares the generation between the dry month of March and the wet month of August.

4.6.3 Energy Purchase Contracts with Renewable Sources

To promote generation from renewable sources, the government created an incentive plan, this produced an avalanche in the implementation of plants with PV and wind technology. The incentive is done in order to identify the energy cost from these sources with the value of the short-term marginal cost, increased by 10% and

Table 6 Comparison between Economic Dispatch and one by PPA

	Dispatch revenue	PPA Dispatch
EnergÑa MWh	168605.6	168605.6
Costo del Despacho US$/semana	14521000	19934000
Costo Marginal US$/MWh	86.12	118.23

added to the above an additional price of USD0.03 cents, resulting in a total cost of USD0.18/kWh equivalent to USD180/MWh. On the other hand, in addition to this high energy cost, the state is forced to buy all the energy produced. This effectively produces a distortion in total generation costs (Table 6). To calculate this distortion, an additional model execution was carried out obtaining the following results:

There is a significant increase between both shipments of USD5,413,000/week, equivalent to USD282, 249, 286/year.

4.6.4 Sensitivity Analysis of Marginal Cost Due to Technology Change

This section analyzes the effect on the marginal cost of generation due to the replacement of an existing plant due to the expiration of the contract for another plant with new generation costs. For this study, the replacement of the LUFUSSA plant was considered, which has a combined capacity of 323.5MW, with the new plant having the same capacity, but different costs of generation. The following graph shows how the marginal cost varies depending on the generation cost of a new technology to replace the LUFUSSA plants, it is observed that if this cost is less than 60 the reduction in the marginal cost is significant; hence, the marginal cost is approaching the cost of generating the new plant.

5 Conclusions

According to the presented methodology outlined in Sect. 3, each of the steps has been positively carried out in the present study. During the methodology monitoring the objectives set out in Sect. 1 have been accomplished. The important conclusions of the study are presented below:

- Power purchase contracts with solar plants produce a strong distortion in the generation system. Especially when compared to an economic firm that follows certain criteria based mainly on a perfect market of supply and demand.
- There is a cost difference between an Economic Dispatch, in which the plants are distributed according to the hourly demand and mainly in the order of their costs, taking into account the capacity restrictions of the generators, the transmission

lines, and other generation systems in which the same criteria of economic generation system are followed; the exception is encounter in ENEE which is forced to distribute all the production of renewable energy plants, especially of the photovoltaic and wind type. The difference is USD5,143,000.00/week equivalent to approximately USD23,200,000/ month and USD282,250,000/year.
- The substitution of generating plants by the expiration of contracts can mean significant reductions in the marginal cost of generation, especially if the plant represents a significant fraction of the generation infrastructure and its production cost is less than the average marginal cost.

As mentioned in the previous sections, this study only intends to make a reduced model of reality in which important conclusions can be obtained. There are great possibilities to expand the horizon of the study, taking into account the complete network of the national systems with all the nodes, all the voltage levels, all the profiles of the generation of plants with renewable energy, power flows, and voltage drops in the lines.

In [33] it is mentioned that short-term dispatch is an optimization task related to highly complex power systems. A series of linear programming techniques, Lagrangian, and Benders decomposition methods have been used to obtain the optimal solution.

By using a simplified network model, its results are non-feasible and distributed due to network limitations by not considering the system in AC. However, simplification not only results in simpler mathematical approaches but a significant reduction in processing time.

Consequently, to carry out a dispatch closer to reality it is not only necessary to increase of amount of information to be introduced into the model, but also the method must be devised in such a way so as to minimize processing time.

The following is a list of tasks that can be additionally performed to obtain broader, more accurate, and more up-to-date results:

(a) Carry out an AC Dispatch in which the voltage drops and losses in the system are considered.
(b) Include all the substations in the system.
(c) Include all transmission and Dispatch lines.
(d) Include a method that streamlines process times, for example, Bender's decomposition techniques.
(e) Create a computer program that facilitates data entry. In this study, to generate the code lines of the current study software was used to generate it. But these were isolated as problem-specific modules. The future proposed program should be more general and whose output should be precisely the source code for executing GAMS suite.
(f) A plan for model execution and periodic publication to monitor the evolution as time passes by.
(g) Use of the model to analyze, study or predict the network behavior for future changes in transmission lines, generation technology, etc.

References

1. H.R. Chamorro, C.A. Ordonez, J.C. Peng, M. Ghandhari, Non-synchronous generation impact on power systems coherency. IET Gener. Transm. Distrib. **10**(10), 2443–2453 (2016). https://doi.org/10.1049/iet-gtd.2015.1233
2. D. Topic, H.R. Chamorro, G. Knezevic, R. Rye, F. Gonzalez-Longatt, V.K. Sood, J. Perko, Analysis of PV systems and charging stations integration into the public lighting infrastructure, in *2019 IEEE PES Innovative Smart Grid Technologies Europe (ISGT-Europe)* (2019), pp. 1–5
3. S. Abu-elzait, R. Parkin, The effect of dispatch strategy on maintaining the economic viability of pv-based microgrids, in *2019 IEEE 46th Photovoltaic Specialists Conference (PVSC)* (2019), pp. 1203–1205
4. J.S. Obando-Ceron, R. Moreno-Chuquen, Impacts of demand response under wind power uncertainty in network-constrained electricity markets, in *IEEE ANDESCON* (2018), pp. 1–5
5. M.H. Albadi, A.S. Al-Hinai, M.J.A. Maharbi, A.M.A. Hosni, M.A.A. Hajri, Economic dispatch of oman's main interconnected system in presence of 500mw solar pv plant in ibri, in *IEEE Jordan International Joint Conference on Electrical Engineering and Information Technology (JEEIT)* (2019), pp. 204–208
6. D. Liu, J. Guo, Y. Huang, W. Wang, P. Wang, A dynamic economic dispatch method of wind integrated power system considering the total probability of wind power, in *2nd IET Renewable Power Generation Conference (RPG 2013)*, (2013), pp. 1–4
7. L. Ran, L. Zhengyu, C. Zhen, Economic dispatch of off-grid photovoltaic generation system with hybrid energy storage, in *2018 2nd IEEE Conference on Energy Internet and Energy System Integration (EI2)* (2018), pp. 1–6
8. C. Rooks, X. Kou, F. Li, Interval optimization for robust economic dispatch in active distribution networks considering uncertainty, in *International Conference on Smart Energy Systems and Technologies (SEST)* (2019), pp. 1–6
9. C. Liu, X. Wang, X. Wu, Y. Zou, H. Zhang, L. Yao, Y. Zhang, Q. Cui, Economic dispatch for microgrid with electric vehicles in plug-in charging and battery swapping modes, in *IEEE PES Asia-Pacific Power and Energy Engineering Conference (APPEEC)* (2016), pp. 1158–1163
10. Y. Peng, S. Gao, B. Xu, Optimal dispatch and techno-economic evaluation of mixed ac and dc distribution networks with high penetration of photovoltaic panels, in *IEEE Asia Power and Energy Engineering Conference (APEEC)* (2019), pp. 258–262
11. J. Zhu, Q. Liu, X. Xiong, J. Ouyang, P. Xuan, P. Xie, J. Zou, Multi-time-scale robust economic dispatching method for the power system with clean energy. J. Eng. **2019**(16), 1377–1381 (2019)
12. N.A. Khan, G.A.S. Sidhu, F. Gao, Optimizing combined emission economic dispatch for solar integrated power systems. IEEE Access **4**, 3340–3348 (2016)
13. A. Soroudi, A. Rabiee, A. Keane, Stochastic real-time scheduling of wind-thermal generation units in an electric utility. IEEE Syst. J. **11**(3), 1622–1631 (2017)
14. M. Abou Houran, W. Chen, M. Zhu, L. Dai, Economic dispatch of grid-connected microgrid for smart building considering the impact of air temperature. IEEE Access **7**, 70332–70342 (2019)
15. Y. Ying, Y. Wu, Y. Su, R. Fu, X. Liang, H. Xu, Dispatching approach for active distribution network considering pv generation reliability and load predicting interval. J. Eng. **2017**(13), 2433–2437 (2017)
16. F. Conte, F. D'Agostino, P. Pongiglione, M. Saviozzi, F. Silvestro, Mixed-integer algorithm for optimal dispatch of integrated pv-storage systems. IEEE Trans. Ind. Appl. **55**(1), 238–247 (2019)
17. H. Wang, S. Wang, Q. Zhao, J. Wang, Bi-level optimisation dispatch method for photovoltaic hosting capacity enhancement of distribution buses. IET Gener. Transm. Distrib. **13**(23), 5413–5422 (2019)
18. Z. Zhao, K. Wang, G. Li, X. Jiang, X. Wang, Economic dispatch of distribution network with inn for electric vehicles and photovoltaic. J. Eng. **2019**(16), 2864–2868 (2019)

19. T. Masuta, T. Oozeki, J.G. da Silver Fonseca, A. Murata, Evaluation of economic-load dispatching control based on forecasted photovoltaic power output, in *ISGT 2014* (2014), pp. 1–5
20. W.C. Flores, P. Meraz, J. Berrios, D. Melara, C. Barahona, W. Sifuentes, The solar eclipse of august 21, 2017 in honduras: Evidence of the impact on the power system operation, in *2018 IEEE PES Transmission Distribution Conference and Exhibition—Latin America (T D-LA)* (2018), pp. 1–5
21. W.C. Flores, Analysis of regulatory framework of electric power market in Honduras: promising and essential changes **20**(1), pp. 46–51. https://doi.org/10.1016/j.jup.2011.11.006, http://www.sciencedirect.com/science/article/pii/S0957178711000816
22. W.C. Flores, Some issues related to the regulatory framework and organizational structure of the central american electricity market, in *2016 IEEE PES Transmission Distribution Conference and Exposition-Latin America (PES T D-LA)* (2016), pp. 1–5
23. E.N.L. Canté, Despacho econÓmico de carga considerando restricciones en la red de transporte con el uso de tócnicas de programaciÓn lineal (2005), 143p
24. A.J. Wood, B.F. Wollenberg, G.B. Sheblé, *Power Generation, Operation, and Control*, 3rd edn. (Wiley-Interscience, 2013)
25. F.S. Hillier, G.J. Lieberman, *Investigacion de Operaciones* (McGraw-Hill Companies, 2002)
26. R.E. Gomory, Outline of an algorithm for integer solutions to linear programs **64**(5), pp. 275–278. https://projecteuclid.org/euclid.bams/1183522679
27. E.C. Ron, A.J.C. Navarro, P.P. Tercero, Formulación Y Resolución De Modelos De Programación Matemática En Ingeniería Y Ciencia, Universidad De Castilla-La Mancha
28. R. Rahmaniani, T.G. Crainic, M. Gendreau, W. Rei, The Benders decomposition algorithm: a literature review **259**(3), pp. 801–817. https://doi.org/10.1016/j.ejor.2016.12.005, http://www.sciencedirect.com/science/article/pii/S0377221716310244
29. M.S. Bazaraa, H.D. Sherali, C.M. Shetty, *Nonlinear Programming: Theory and Algorithms*, 3rd edn. (Wiley-Interscience, 2013)
30. D.P. Bertsekas, *Nonlinear Programming*, 3rd edn. (Athena Scientific, 1994)
31. D.G. Luenberger, Y. Ye, *Linear and Nonlinear Programming*, 4th edn. (Springer, 1984)
32. A. Soroudi, *Power System Optimization Modeling in GAMS* (Springer International Publishing, 2017) https://doi.org/10.1007/978-3-319-62350-4, https://www.springer.com/gp/book/9783319623498
33. W.S. Sifuentes, A. Vargas, Hydrothermal scheduling using benders decomposition: accelerating techniques. IEEE Trans. Power Syst. **22**(3), 1351–1359 (2007)

Converter/Inverter Topologies for Standalone and Grid-Connected PV Systems

Sudhakar Babu Thanikanti, Dalia Yousri, Dalia Allam, M. B. Etebia, and Karthik Balasubramanian

1 Introduction

In today's world, energy is extremely costly; hence saving energy is extremely vital. Attempts must be made to save energy and make it available to mankind. Renewable sources such as solar, wind, tidal, etc., contribute effectively to curbing the inefficient use of power. Renewable energy is environmentally friendly and hence, significant countries today are utilizing power generated from renewable energy sources. According to the World Energy Outlook 2018 report, by 2040 photovoltaic (PV) will become more relevant, with a global capacity to produce more electricity than any other process, except natural gas [1]. Currently, Solar PV usage is high in demand and considered economic friendly.

Tracking of MPPT efficiently in Solar PV remains a challenge for scientists and engineers due to its unique characteristics in nature [2]. To alleviate this issue, electronic power converters serve a vital role in regulating and extracting maximum

S. B. Thanikanti (✉)
Department of Electrical and Electronics Engineering, Chaitanya Bharathi Institute of Technology (CBIT), Hyderabad, India
e-mail: sudhakarbabu66@gmail.com

D. Yousri · D. Allam · M. B. Etebia
Faculty of Engineering, Electrical Engineering Department, Fayoum University, Fayoum, Egypt
e-mail: day01@fayoum.edu.eg

D. Allam
e-mail: dfallam@yahoo.com

M. B. Etebia
e-mail: meteiba@gmail.com

K. Balasubramanian
Offshore Technology Development Pte. Ltd, Keppel Offshore and Marine Ltd, 55 Gul Road, Singapore, Singapore
e-mail: karth_92@yahoo.com

V. K. Singh et al. (eds.), *Renewable Energy and Future Power Systems*, Energy Systems in Electrical Engineering, https://doi.org/10.1007/978-981-33-6753-1_2

power. Toward this end, the authors have designed a boost converter using an optimization algorithm to track the maximum power [3]. Application of maximum power extraction techniques (MPPT), including those with numerous converter topologies, is presented in [4]. A detailed analysis has been performed on the existing converter and inverter topologies in this review paper.

Common applications of converters/inverters are UPS power supplies and modern electrical drives [5]. Single-stage converters are effective and efficient, but one of the common limitations of these types of converters is the startup operation, which requires attention. Also, the cost of the system increases due to the requirement of feedback controllers in power electronic converters [6, 7]. To effectively overcome the issue of leakage current in non-isolated PV grid, various circuit topologies at the inverter side have been developed. PV systems can be broadly segregated as follows (1) standalone, (2) grid connected.

1.1 Grid-Connected Solar PV

The expansion of smart grid communication, most renewable energy plants are developing significantly [8]. Grid-connected systems are also designed for the charging operations for electric vehicles [9]. Transformers are primarily used for conversion of power in a grid, where the voltage and frequency are supported by the utility grid. AC-AC conversion is most commonly performed with the support of the transformer [10]; DC-DC conversion is a more convoluted process due to its unstable nature. Reduction of voltage can be performed with power diodes and voltage bridges, but these are usually ineffective. Another well-known fact is that the voltage of a battery is unstable due to its discharging property. Hence, to eradicate the aforementioned limitations, the best-suited method to perform DC-DC conversion is with the help of a DC-DC converter. The schematic of the grid-connected system is shown in Fig. 1.

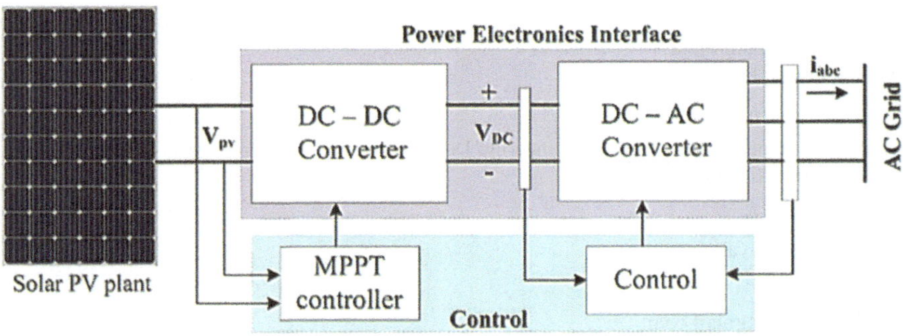

Fig. 1 Schematic of the grid-connected PV system

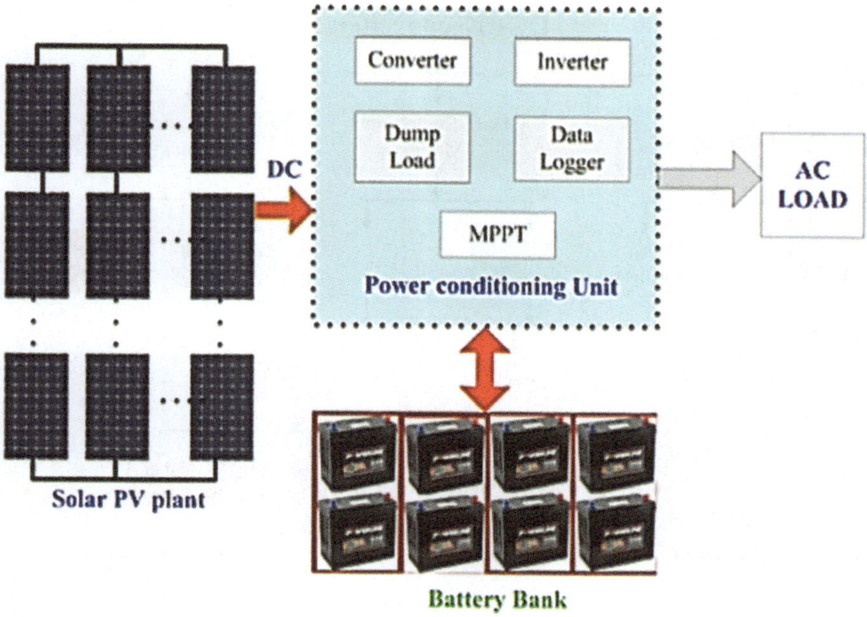

Fig. 2 Standalone PV network scheme

1.2 Standalone PV Systems

The concept of standalone systems is best explained with the inverter where DC current is drawn from batteries. The size of the battery unit decides the lifetime of the PV system [6, 11].

The major utilizations of converters are for increases or reductions in voltage, which are performed by boost and buck converters, respectively [12, 13]. Figure 2 represents the standalone PV network scheme.

The various converter topologies work based on maximum power extraction techniques are presented in Sect. 2. Inverter topologies in grid applications along with its comparative studies are detailed in Sect. 3. Section 4 deals with various widely used standards for the grid connected PV system. Recent advancements and challenges to design converter/inverter topologies are discussed in Sect. 5. Directions for future research are detailed in Sect. 6. Section 7 highlights the conclusions.

2 MPPT-Based DC-DC Converter Topologies

DC-DC converters are noted as one of the essential components in solar PV systems as they enhance power generation [14]. The three modes of converter operation are: (1) linear mode, (2) hard switching, and (3) soft-switching mode. Classification of

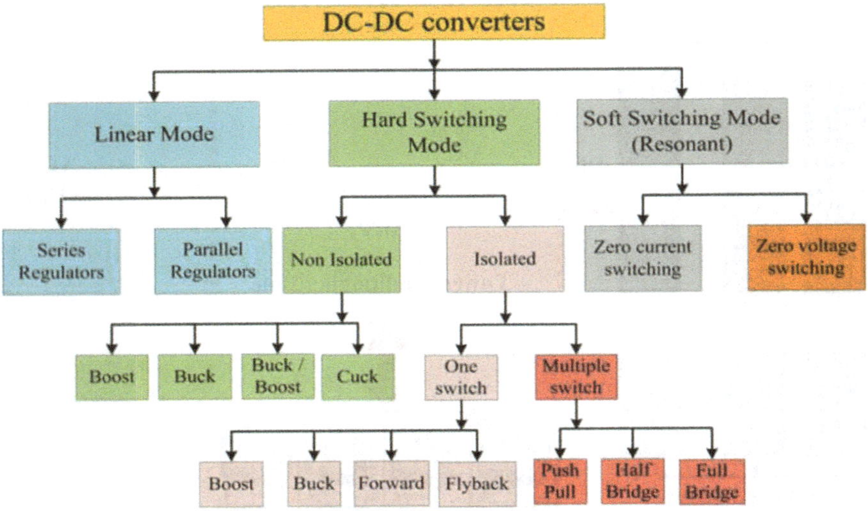

Fig. 3 Classification of converter topologies

converters is shown in Fig. 3. Linear mode technology exhibit features like reduced noise, simple design, and quick response to the finest regulation. However, this type of converter performs less efficiently due to the dissipation of power during operation.

Switching mode technology is categorized into two types, isolated and non-isolated. The power converters operate with a voltage level between 100 V and 600 V, which requires isolation for safety measures [15, 16]. These operate with high frequency transformers; therefore, single-switch converters are not preferable for high voltage ratings. To address this issue, DC-DC converters with multiple switches were introduced, such as full/half-bridge and push-pull converters. Further, the hard-switching converter leads to lower efficiency due to dissipation of more power. In addition, this type of switching causes high electromagnetic interference, increases weight, is larger, and operates with limited frequency. To alleviate the drawbacks of these methods, soft-switching converters were developed by the authors [17, 18]. Numerous studies have been carried out by researchers on resonant converters such as parallel and series converters as presented in [17]. In this article [17], the authors focused on the characteristic behavior of parallel and series resonant converters and low-order resonant converters. Resonant converters with multiple elements have been presented by the authors [19]. Comparative studies between isolated and non-isolated converters have also been presented by the authors [20]. This article addresses DC-DC converters for high voltage, which are ideal for PV applications [21]. Soft-switching and resonant converters classifications are seen in Fig. 4.

Detailed discussions of these categories of converter topologies are reviewed in this article. Features, limitations, and applications of various converter topologies are summarized in Table 1. After thorough analysis, it was noticed that the boost converter yielded high efficiency at a lower duty, while the buck converter yielded

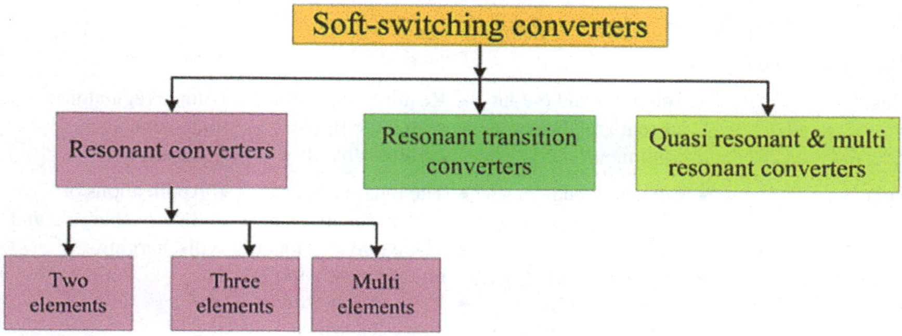

Fig. 4 Classification of resonant converters

Table 1 Characteristics of various converter topologies

Type of converter	Advantages	Drawbacks	Applications
Buck	• The absence of control complexity and high efficiency • Inductor helps to limit the *di/dt* of the load current • The voltage ripple is extremely minimum	• The power factor is extremely poor • Flux density is high at the inductor • The buck converter requires the additional protection circuit	• Major applications are in independent power supplies • LED arrays • Point of load converters in servers
Boost	• Ability to boost voltage at less component count • Provides continuous current which is more desirable for PV and battery sources	• The output capacitor is having discontinuous charging current • Increase capacitor size and issues existent in EMI • Due to the high duty cycle it exhibits poor efficiency	• Mainly used in the Automotive industry
Buck boost	• PF correction factor is high • Provides inverted output • High efficiency • Extremely good short circuit withstanding capacity	• Voltage stress on switching device in the chopper is high	• Adaptive control applications • Power amplifier applications
SEPIC	• It provides positive voltage • This converter also eliminates the fluctuating DC voltage	• The design of this converter is extremely convoluted	• NiMH chargers • DC power supplies with a wide range of input voltages

(continued)

Table 1 (continued)

Type of converter	Advantages	Drawbacks	Applications
Cuk	• The input and output current are always continuous	• Requires capacitor with large ripple current capability	Voltage regulator in solar-wind energy applications
Zeta	• Efficiency and voltage gain are very good • The output voltage is positive in reference to the ground making the sensing circuit simple	• The input current is discontinuous • The passive element requirement is more • It is a fourth order converter, which makes the control difficult	• Applications of Battery charging and discharging
Landsman	• It has smooth transient response and good dynamic output voltage with reduced ripple and settling time	• This converter fails to protect against overload and short-circuit current • It is not suitable for discontinuous mode of operation • Difficulty in handling of magnetic inrush current • Core saturation problem	• Widely used in speed control of induction motors • Power factor correction
Forward	• Duty cycle is limited in range between 27% to 55%. • Multiple outputs can be regulated/quasi regulated	• Performs poor transient response • Requirement of maximum duty cycle clamp is required	• Widely used in medium and lower power applications
Resonant push-pull	• Minimum output current ripple • Best suitable for lower input voltage	• High voltage on primary switches • Transformer flux walking	• Power rating < 500 W, telecom or low Vin (<200 V)
Fly back	• Due to the presence of a transformer, it has more flexibility in designing	• Duty cycle is limited to 50%	• The main applications of the fly back converter are lasers

high efficiency at high duty. Moreover, Buck–Boost, Zeta, and Landsman converters provided higher efficiency than other converters. Further, diode losses in SEPIC and Fly back converters were far lower compared to those of other converter topologies.

DC-DC converter strategies used for the application of MPPT are reviewed in the following subsections. Figure 5 shows the MPPT controller with converter.

Widely used notable converters are presented in Table 2. In this table, a schematic of the converter circuit, characteristic waveforms, and important formulas in the building of the converter is presented.

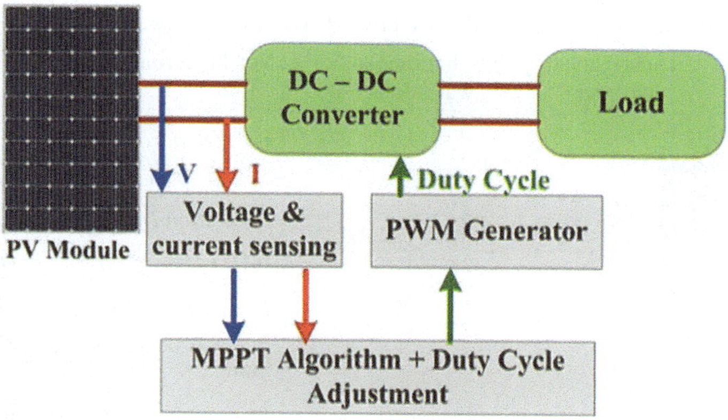

Fig. 5 MPPT controller with converter

Regarding the PV system application, a comparison table of various noticeable parameters between converters is presented in Table 3. After application of MPPT using different converter topologies, a comparative study of efficiency, tracking time, and other important parameters were conducted, with results presented in Table 4, which can be helpful for researchers seeking better converter topology with minimal losses for PV applications.

In addition, with successful implantation of various converter topologies for tracking maximum power, observed points from the literature are listed in Table 5. Table 5 describes the various MPPT techniques used by the authors with different converter topologies and associated remarks.

The authors have introduced specific MPPT algorithms to control topologies for maximum power converters. The regional MPPT approaches have a broader range of benefits than traditional methods across the various MPPT technologies. These were highly robust, effective to operate under various types of irradiation conditions, performed wide searches for optimal solutions, exhibited faster convergence, and were more reliable. Conventional methods were limited by their failure in tracking maximum power in cases of sudden irradiation changes. In addition, these methods required additional sensors to track maximum power. To overcome these drawbacks, global MPPT methods were introduced. These methods have a wide range of advantages such as effectiveness in various types of irradiation changes, wide search possibilities, effective use of search space, and highly effective handling of non-linearity [4].

Table 2 Schematics and characteristic waveforms of converter topologies

Name of converter	Schematic of converter	Characteristic waveforms	Notable formulas
Buck [71, 72] Figure 6	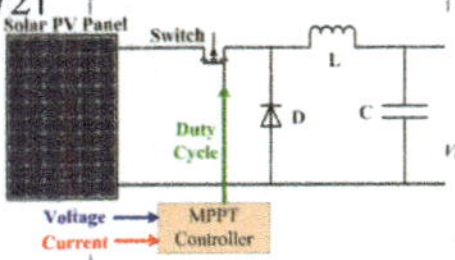	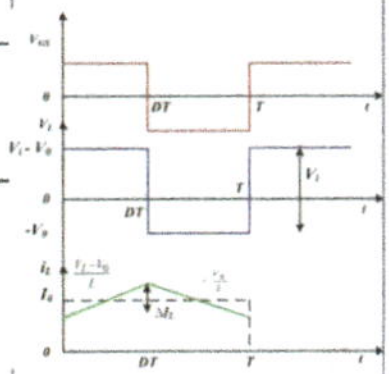	$Voltage\ across\ Inductor = V_L = -V_0 = L\frac{dI_L}{dt}$ $Maximum\ Inductor\ current = I_L(DT) = \frac{(V_I - V_0)DT}{L} + I_L(0)$ $The\ value\ of\ Indcutcance = L > L_{min} = \frac{R_L max(1 - D_{min})}{2f_s}$ $The\ value\ of\ capacitor = C_{min} = \frac{\Delta I_L max}{8 f_s V_{Cpp}} = \frac{(1-D_{min})V_0}{8 f_s^2 L V_{cpp}}$
Boost [73] Figure 7	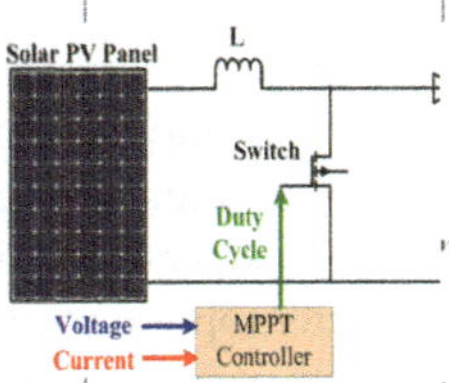	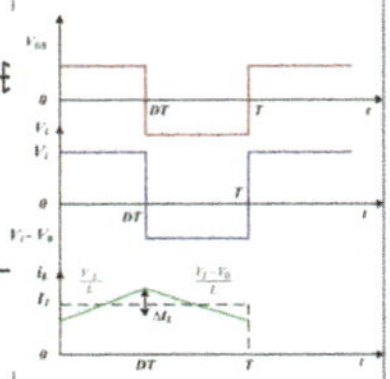	Output voltage $= V_{out} = \frac{V_{in}}{1-D}$ Capacitance $= C = \frac{DV_0}{R\Delta V_0 f}$ Inductor value $= L = \frac{D(1-D)^2 R}{2f}$
Buck boost [74] Figure 8	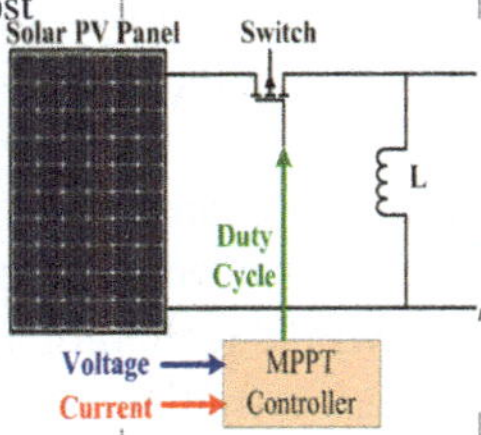	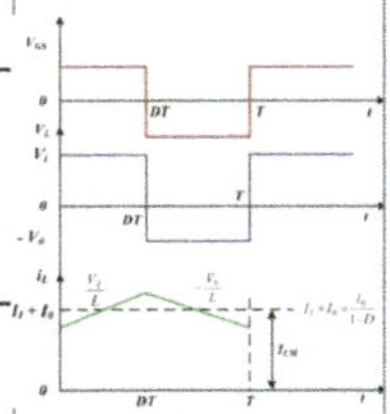	Output voltage $= V_o = \frac{D}{1-D} V_d$ Output current $= I_o = \frac{1-D}{D} I_d$ Inductor value $= L = \frac{D(1-D)^2 R}{2f}$
SEPIC [75–77] Figure 9	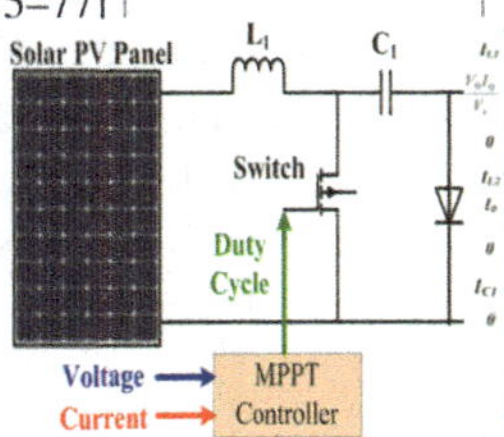	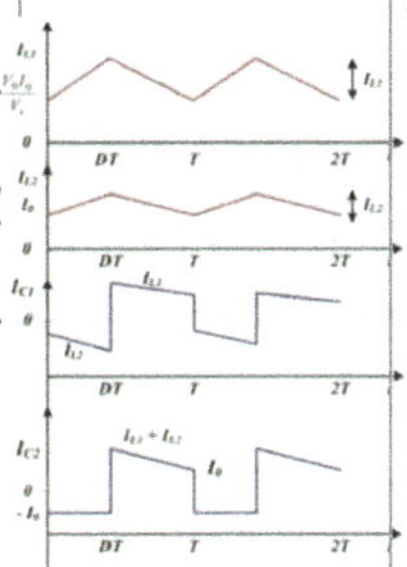	Output voltage $= V_{out} = \frac{DV_{in}}{1-D}$ Input inductance $= L_1 = \frac{D \times V_{pv}}{f \times \Delta I_{L1}}$ Output inductance $= L_2 = \frac{(1-D) \times V_{DC}}{f \times \Delta I_{L2}}$ Intermediate capacitance $= C_1 = \frac{D \times I_{out}}{f \times \Delta V_{c1}}$

(continued)

Table 2 (continued)

Name of converter	Schematic of converter	Characteristic waveforms	Notable formulas
Cuk [78–80] Figure 10	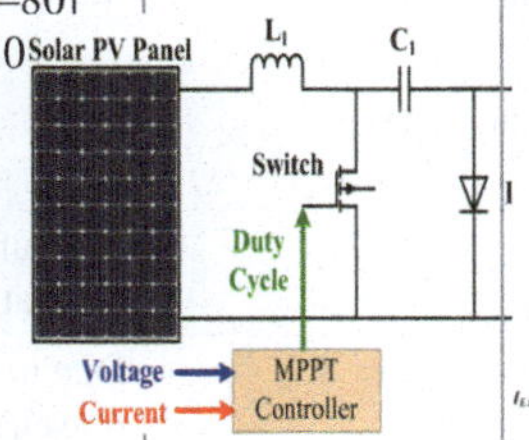	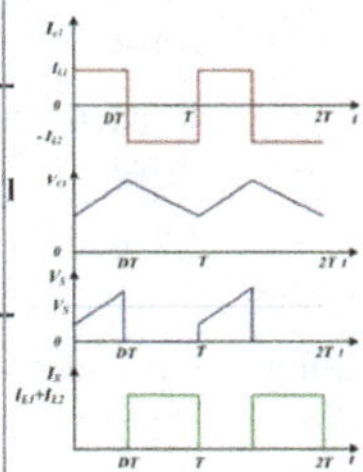	Output voltage = $\frac{\Delta V_0}{V_0} = \frac{1-D}{8L_2C_2f^2}$
Zeta [81] Figure 11	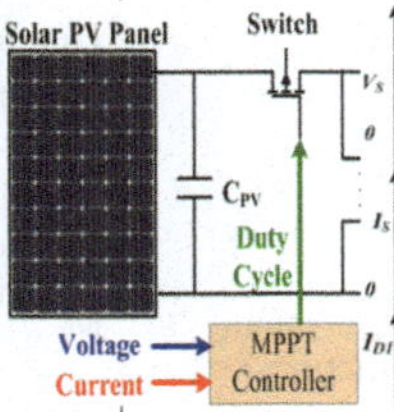	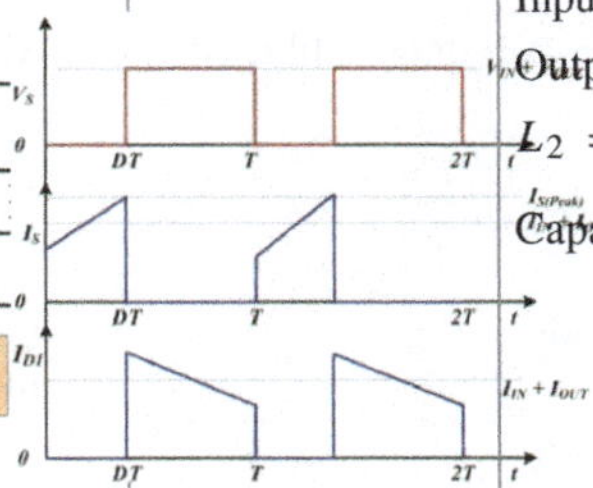	Input Inductance = $L_1 = \frac{D\times V_{pv}}{f\times\Delta I_{L1}}$ Output inductance = $L_2 = \frac{(1-D)\times V_{DC}}{f\times\Delta I_{L2}}$ Capacitance = $C_1 = \frac{D\times I_{out}}{f\times\Delta V_{c1}}$
Landsman [82] Figure 12	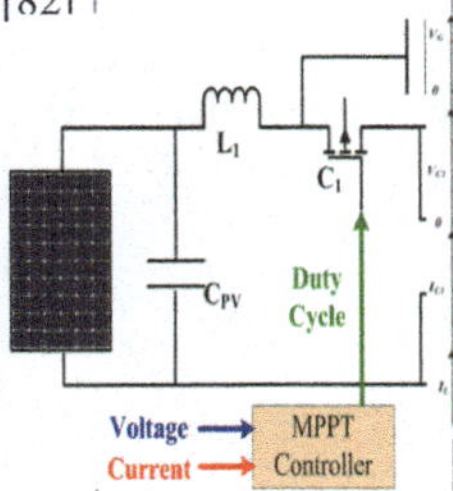	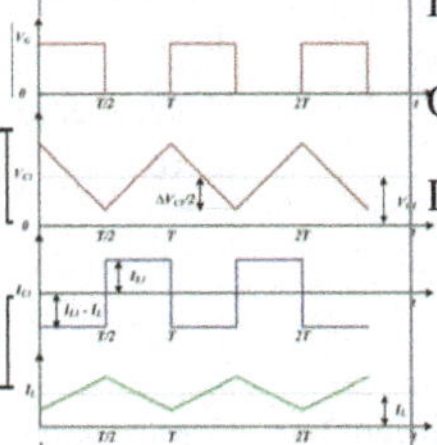	Input inductance = $L_2 = \frac{(1-D)\times V_{DC}}{f\times\Delta I_{L2}}$ Output inductance = $L_1 = \frac{D\times V_{pv}}{f\times\Delta I_{L1}}$ DC link capacitance = $C_1 = \frac{D\times I_0}{f\times\Delta V_{c1}}$
Forward [83] Figure 13	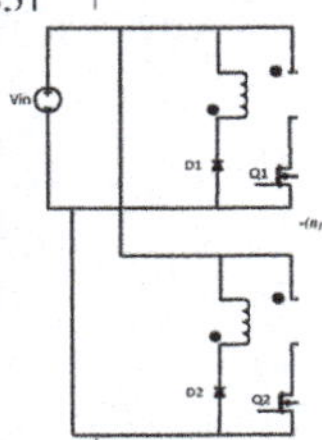	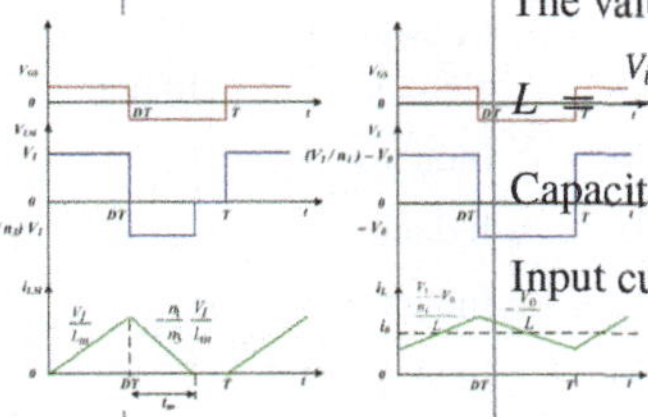	The value of inductance = $L = \frac{V_{in}\times\frac{N_s}{N_p}-V_{out}}{F}\times\frac{D}{\Delta i}$ Capacitance = $C = \frac{\Delta i}{8F\Delta V}$ Input current = $I_{in} = \frac{I_{out}}{1-D}\times\frac{N_s}{N_p}$

Table 3 Comparison of different converters based on noticeable parameters

Converter	Nature of polarity	I/P Current	Switch drive	Competence	Price
Buck	Non-Inverting	Pulsing	Floated gate drive	H	M
Boost	Non-inverting	Pulsing	Floated gate drive	H	M
Buck-Boost	Inverting	Pulsating	Floated	L	Medium due to float drive
SEPIC	Non-inverting	Non-Pulsating	Grounded	M	Due to the presence of block capacitor, the cost is medium
Cuk	Non-Inverting	Non-Pulsating	Floated	M	Due to the presence of block capacitor, the cost is medium
Zeta	Non-inverting	Steady	–	Less than SEPIC	M
Landsman	Inverting	Steady	–	Less than SEPIC	M
Forward	Inverting	Pulsating	Floated gate drive	M	H
Resonant Push-pull	Inverting	Steady	Floated gate drive	Higher than buck–boost	H
Fly back	Inverting	–	–	M	–

Where M—Medium, H—High, L—Low

Table 4 Comparative analysis of converters based on its design

Type of converter	Comparative parameters					
	Hardware complexity	Cost	Energy transferring elements	Average tracking efficiency	Efficiency of converter	Switching stress
Buck	M	MM	Inductor	M	H	L
Boost	M	ML	Inductor	L	H	L
Buck–Boost	M	ML	Inductor	H	M	L
SEPIC	H	MM	Inductor and capacitor	H	M	M
Fly back	H	ML	Transformer	H	M	L
References	Refs. [84, 85]	Refs. [86–88]	Refs. [87–90]	Refs. [55, 91–93]	Refs. [85, 94–97]	Refs. [85–87]

*Where L—Low, M—Medium, H—High

Note: L, M, H represents Low, Medium, and High respectively.

Table 5 Analysis of various converter topologies used for the application of MPPT

Reference	Converter topology	Considered MPPT method	Remarks
Haroun et al. [98]	Cascade connection of boost converter with sliding mode control	Extremum seeking algorithm	• Requires additional auxiliary circuits for sliding mode control • Lower conversion efficiency
Tamyurek and Kirimer [99]	Three cells interleaved flyback converter	Perturb and Observe	• Formation of transient voltage due to transformer leakage inductance
El Khateb et al. [100]	Fuzzy logic controller based SEPIC converter	Fuzzy logic controller	• Maximum static gain is obtained when the duty cycle values are optimal
Quamruzzaman et al. [101]	SEPIC with coupled inductor	Perturb and Observe	• Non-zero DC parasite capacitance induction voltage mismatch can induce ripples
Liao et al. [102]	Four switch buck–boost converter	Perturb and Observe	• It requires compensator
Babu et al. [3]	Boost converter	Modified Particle swarm optimization	• Output power reduces with consideration of light loads
Ishaque and Salam [103]	Buck–Boost converter	Particle swarm optimization	• Variations in irradiation causes the change in operating modes of this algorithm
Gounden et al. [104]	3 phase converter	Fuzzy based MPPT algorithm	• Exceptional technique of employing fuzzy to an inverter • The MPP is achieved by operating the controller in optimal firing angle
Liu et al. [105]	Push-Pull converter	Incremental conductance algorithm	• As improvement to traditional IC technique, a variable step size IC technique is proposed

3 Inverter Topologies for PV System

Inverters are referred to as the brain of the renewable energy system as they provide the load requirements [22]. They are interfaced with the output of the converters. Consequently, inverters can be widely used in grid-connected systems and also applied to real time scenarios. For PV systems, inverters may be included in several schemes, such as the grid-connected string inverter, grid-connected central inverter, micro-inverter, multilevel inverter topology, grid-connected inverter, transformer less inverter, grid-connected isolated inverter, and multistage, isolated micro-inverter. Each type has its own merits, drawbacks, and applications as summarized in Table 5. As detailed in Table 6, these inverters with the widest range of applications are the micro-inverter, multilevel inverter topologies, and other grid-connected inverter topologies. It can also be inferred from Table 6 that the inverter with the highest efficiency is the grid-connected inverter topology, with a special mention offered to the grid-connected transformer less inverter and its efficiency of 98% compared to all other conventional inverters. The investment required for the grid-connected string central inverter is much lower, and it therefore provides a huge advantage commercially. Micro-inverters have the ability to generate a high panel MPPT energy yield. It must be also noted that those inverters that are most suitable for residential and commercial purposes are based on the rating of the system. Recent trends in inverter topology [23] and inverter topologies used for the grid-connected PV system has been discussed by the authors [24].

3.1 String Inverter

Figure 14 shows the typical grid-connected string inverter system. The string inverter provides high accurate control on tracking MPP during partial shading [25]. PV plants guarantee effective power production. Hence, utilization of inverters is highly mandatory in large-scale and medium-scale industries due to effective and efficient power generation, thereby improving productivity. Each string has its own MPPT controller; central inverters have DC-DC converters at their second stage [26]. One of the major limitations of the string inverter is that the most extreme yield current of the string is just as high as the weakest irradiated solar PV panel in the string. The authors have detailed various kinds of grid-connected inverters [27].

3.2 Multi String Inverter

This type of inverter is mainly produced for better MPPT accuracy and adaptability. The multi-string inverter is illustrated in Fig. 15.

3.3 Central Inverter

Central inverter serves multiple series of panels; hence, these types of inverters are huge in size. There exists another form of string inverter in which the inverters are

Table 6 Characteristics of various inverter topologies

Type of inverter	Advantages	Disadvantages	Applications
Grid-connected string inverter	• Only one string connected inverter is required • Highly efficient and robust in nature	• If a failure occurs at one point, the complete PV array operation will collapse • It is not very expandable	• Mainly applied in grid-connected PV systems • Mostly used for high rated power plants 2–30 kW
Grid-connected central inverter	• Low capital price per watt • High efficiency	• High loss of control due to concentrated MPPT • Non flexibility design	• These inverters can be used even for higher ratings of power • Most suitable for utility level PV systems and for large commercial applications • The initial rating of these inverters will be around 200 kW
Micro inverter	• Power enhancement is high • Exhibits panel level MPPT • Simple design systems due to plug & play characteristics	• Extremely high price • Many inverters on the roof due to number of solar panels • Increased complexity in installation	• Widely used for residential and small-scale business applications • These can be used as an alternate to string inverter to maximize energy production
Multilevel inverter topology	• Multilevel inverters are active filtered in nature with the ability to double the total number output voltage levels	• The maximum output voltage is half of the input DC voltage	• Induction motor control
Grid-connected transformer less inverters	• These inverters give high efficiency 98% • These are less in weight, more compact, and more affordable due to absence of transformer • In these inverters electronic switching is used, thus reduces the amount of heat and ˜hum' generated by the unit	• Leakage current issue • Needing additional capacitors and dead time, which causes rise the losses and output current distortion respectively • Greater in conduction • Overheating and chances of may occur due to missing line frequency of transformer	• Mainly used in applications to achieve unity power factor

(continued)

Table 6 (continued)

Type of inverter	Advantages	Disadvantages	Applications
Grid-connected isolated inverters	• These inverters achieve maximum efficiency around 96–96.5%	• These systems contain high frequency transformers, switching losses, and core losses • High cost • Less compact	• Grid-connected PV systems • These types of inverters are widely used in applications where the leakage current should made zero

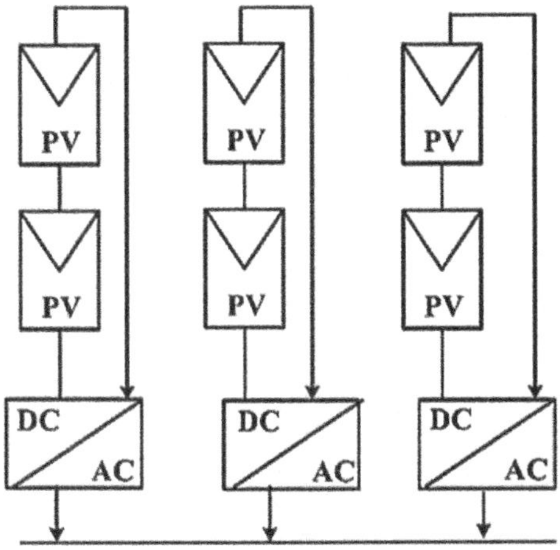

Fig. 14 String Inverter

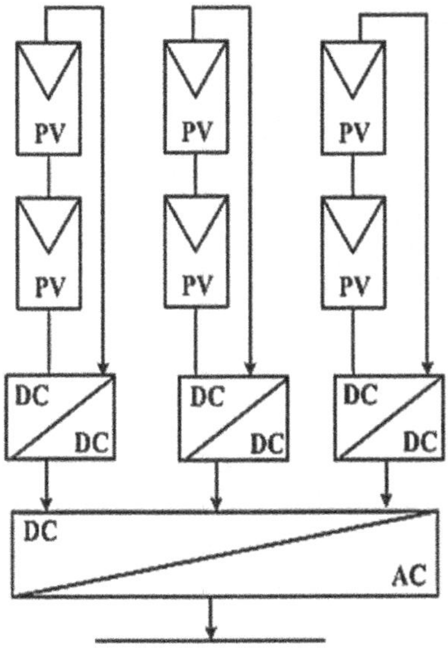

Fig. 15 Multi string Inverter

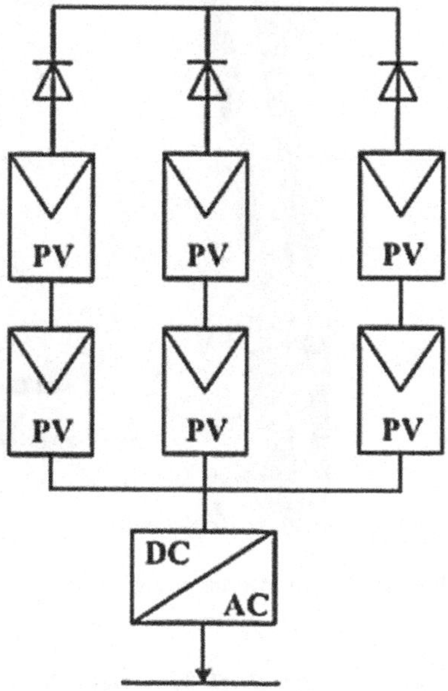

Fig. 16 Grid-connected central inverter

connected to a combiner box [28]. Figure 16 indicates the representation of grid connected central inverter.

3.4 *Micro Inverter*

These types of inverters are connected to each individual PV module. They handle single or multiple panels and are usually connected behind PV modules. Arrangement of the micro-inverter in real time is presented in Fig. 17.

3.4.1 Small Scale PV Inverters

(A) *Standalone inverter*: Standalone systems are preferred in remote areas, in which inverter receives power from charged batteries by PV arrays. Integral battery charges are present in standalone inverters; hence, these types of inverters do not interfere with the utility grid.

(B) *Grid-tie inverter*: The grid tie inverter does not provide backup during outage of utilities.

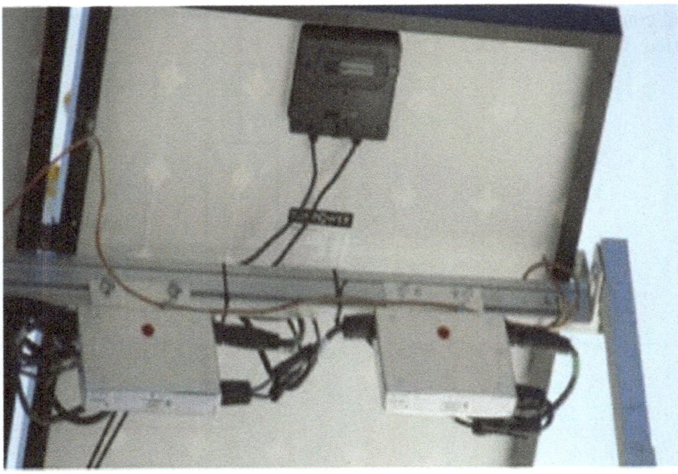

Fig. 17 Micro-inverter arrangement

3.5 Large and Medium Scale PV Inverters

Inverters are the main source of backup power for industries. The following section describes the different topologies of inverters used widely in large and medium-sized PV plants. The authors have previously presented the major types of PV inverters in detail [28].

3.5.1 Multilevel Inverter Topology

Figure 18 shows the representation of n-level multi-inverter. The authors have proposed a device count reduction technique of existing cascaded MLIs to reduce the number of switches [29, 30].

3.6 Grid-Connected Transformer Less Inverters

3.6.1 Two-Stage Topologies

A schematic for the two-stage inverter topology is illustrated in Fig. 19. The high current is pumped to the grid through a PWM controller at the DC-AC stage.

The less value of DC link capacitance (Cp) at high voltage levels is allowed as it enhances the lifespan of the inverter.

The authors have previously discussed a topology of the transformer less inverter as illustrated in Fig. 20 [31, 32]. They have proposed various designs to increase

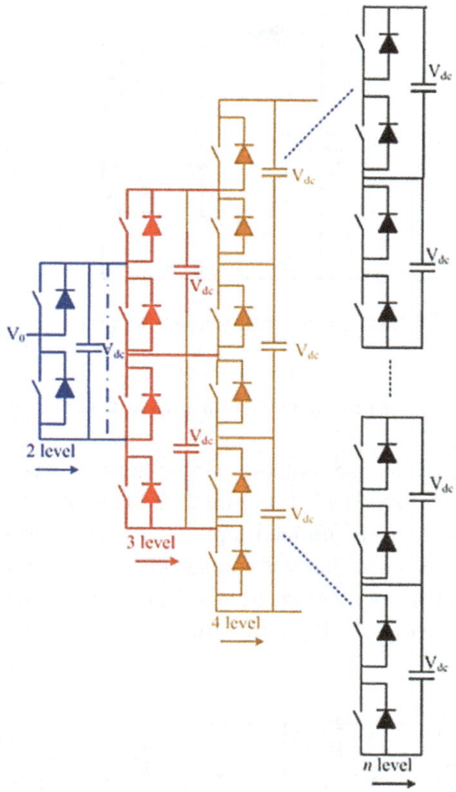

Fig. 18 n-level multilevel inverter topology

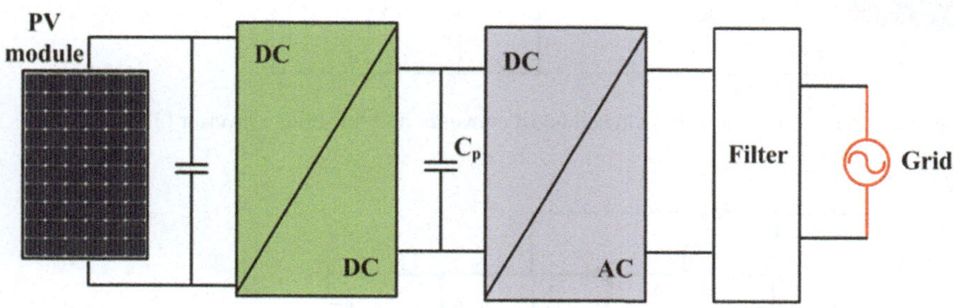

Fig. 19 Two-stage topology of a transformer less inverter

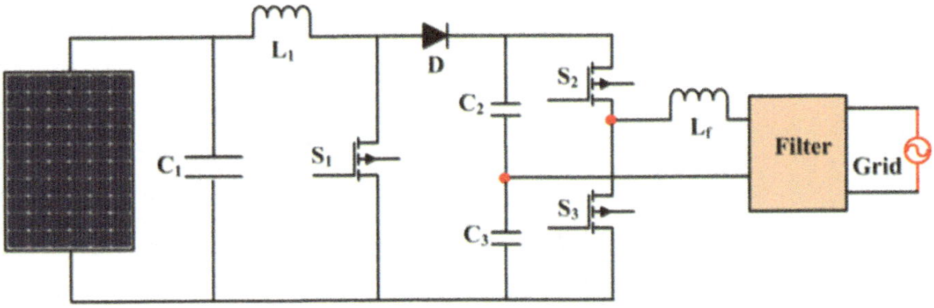

Fig. 20 Boost converter half-bridge transformer less inverter [31]

the productivity. A time period sharing, dual-mode inverter was introduced and is presented in Fig. 21 [31].

The soft-switched resonance structure strengthens converters with dual bridge inverters, as seen in the Fig. 21 [32]. Figure 22 adds a DC-DC transformer flipped power. As in the previously mentioned topologies, the authors have improved the performance of the boost converter with dual bridge converters [32].

Dual-grounded transformer less inverters [33] have been considered for solar PV systems, as presented in Fig. 23. Figure 23 shows the topology in which a portion of

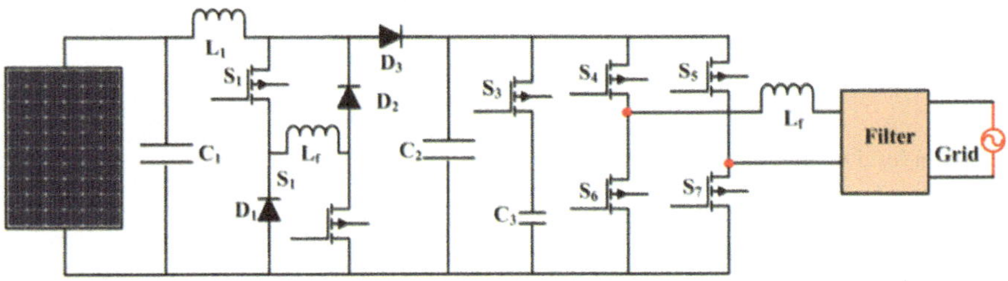

Fig. 21 Parallel resonant soft-switched boost converter and full-bridge inverter [32]

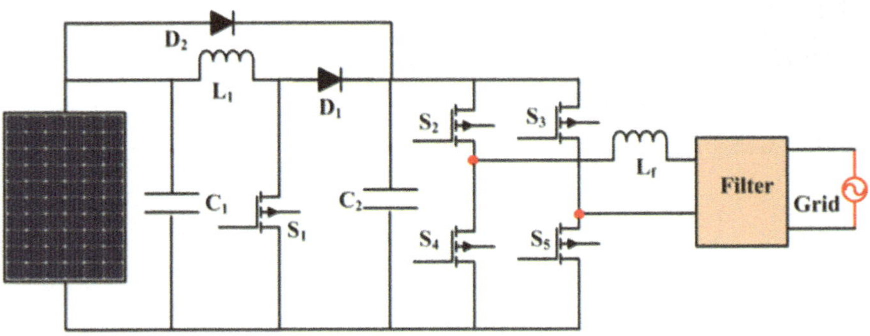

Fig. 22 Full-bridge transformer less inverter [32]

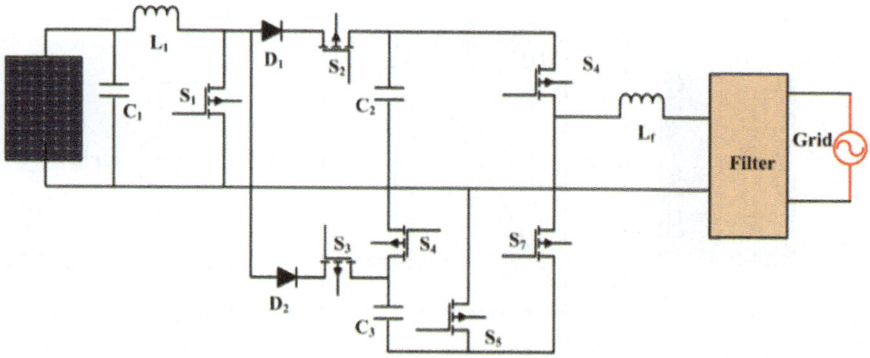

Fig. 23 Boost with feedback integrated and dual-grounded transformer less inverter [33]

Table 7 Analysis on two-stage inverter topologies

Figure No. and Reference	V_{out} (V)	Power rating (W)	Frequency (kHz)	Switching time (ms)	Efficiency (%)
Figure 21 and [106]	200	1600	20	0.05	96.5
Figure 22 and [31]	120	270	30	0.033	–
Figure 23 and [32]	230	980	6.26	0.15	–
Figure 24 and [33]	210	1500	20	0.05	94

the switches is shared by the two phases. Total Harmonic Distortion <3%, and high productivity, >94%, accounted for main elements of the two-stage topology inverters as shown in Table 5. Table 7 offers a thorough overview of the two-stage topologies of inverters discussed in traditional papers [34, 35]. From Table 7 the authors concluded that the inverter presented in Fig. 23 was the most efficient compared to all other existing inverters [33].

3.6.2 Single-Stage Topologies

The main goals of the development of effective inverters by scientists and researchers have been to reduce cost, and, thereby, increase power density [36]. Figure 24 represents the single-stage topology of the step-up transformer less inverter that is widely used in most industries [37].

Figure 25 illustrates the universal, single-stage, grid-connected inverter. In this type of inverter, all converted topologies are incorporated by the multistage inverter topologies. Figures 26 and 27 present the integrated boost converter and buck–boost

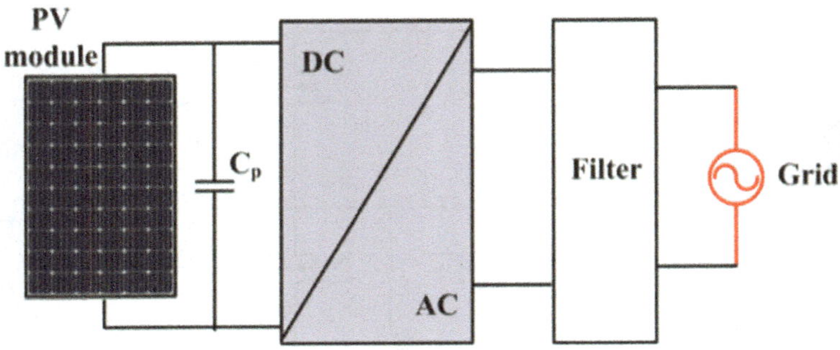

Fig. 24 Single-stage topology for an AC module

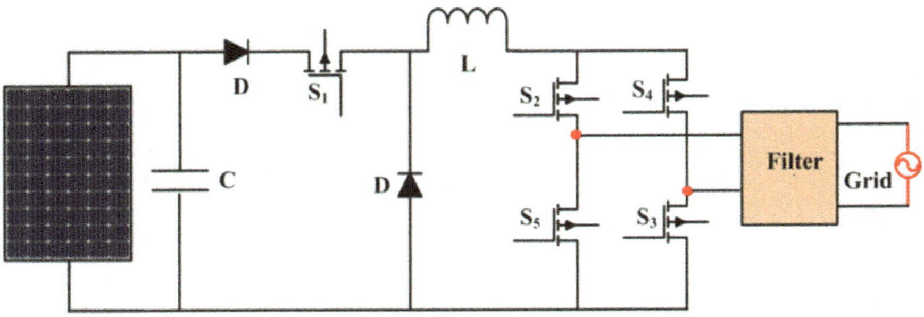

Fig. 25 Universal single-stage grid-connected inverter [66]

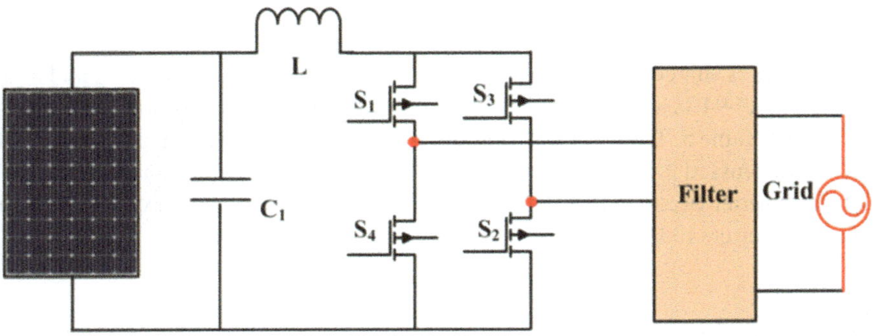

Fig. 26 Integrated boost inverter [38]

converter, respectively [38]. The same topology, which achieved less voltage, thus adding to the THD, was broken down for independent applications [39].

Liang et al. [39] shows the improved zero crossing boost converter is presented in Fig. 28. [40] presents a schematic for the same is shown in Fig. 29. In this system, the DC-biased sinusoidal differential setup was considered for the buck–boost inverter as shown in Fig. 30 and discussed by the authors [41]. Nevertheless, for every positive

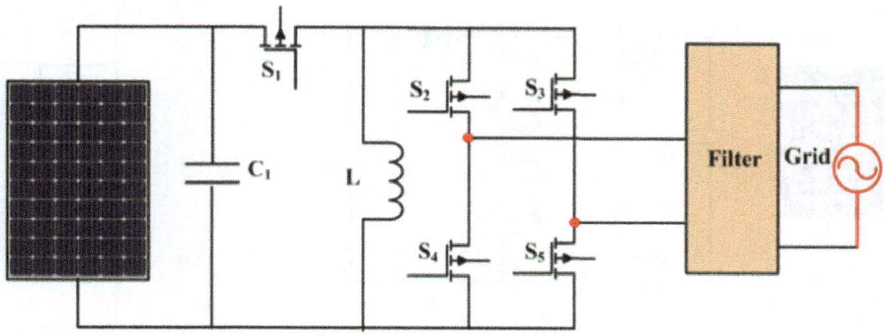

Fig. 27 Integrated buck–boost inverter [38]

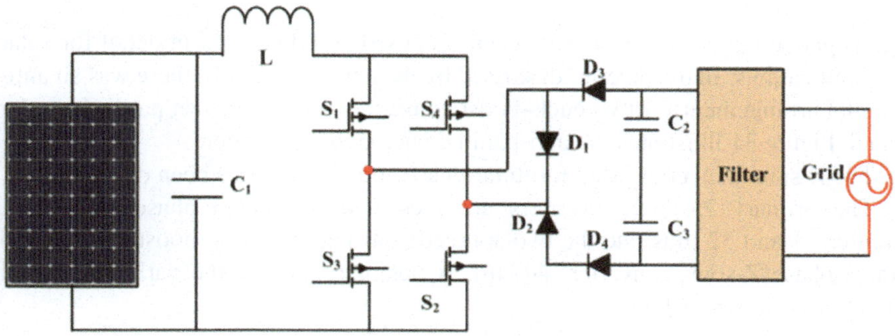

Fig. 28 Improved zero crossing boost converter [39]

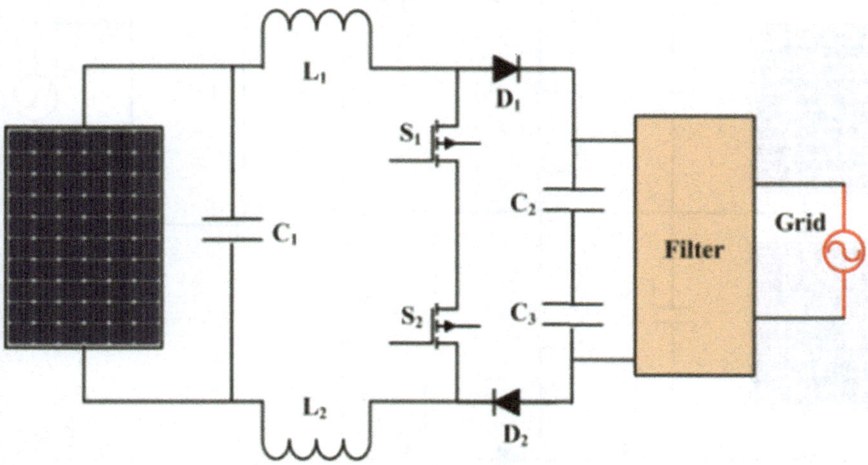

Fig. 29 Differential boost inverter [40]

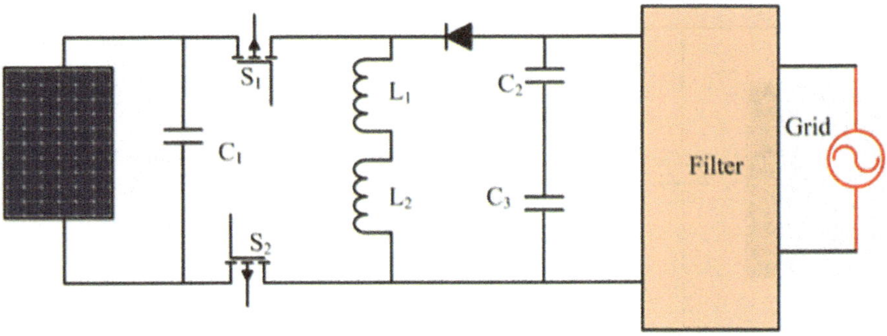

Fig. 30 Differential buck–boost inverter [41]

and negative lattice, voltages have been proposed and discussed in detail for solar PV applications. In the inverter discussed by the authors [42, 43], there was an anti-parallel arrangement for two buck–boost converters with their own particular solar panel. Figure 31 illustrates the anti-parallel buck boost converter.

The Z-source inverter, used for three-phase applications, has been detailed [44].

The primary Z-source inverters are presented for single phase topologies. Figures 31 and 32 illustrate the two-sourced, anti-parallel buck–boost inverter and single-phase Z-source inverter [44–46]. A detailed study of the various types of inverters is shown in Table 6.

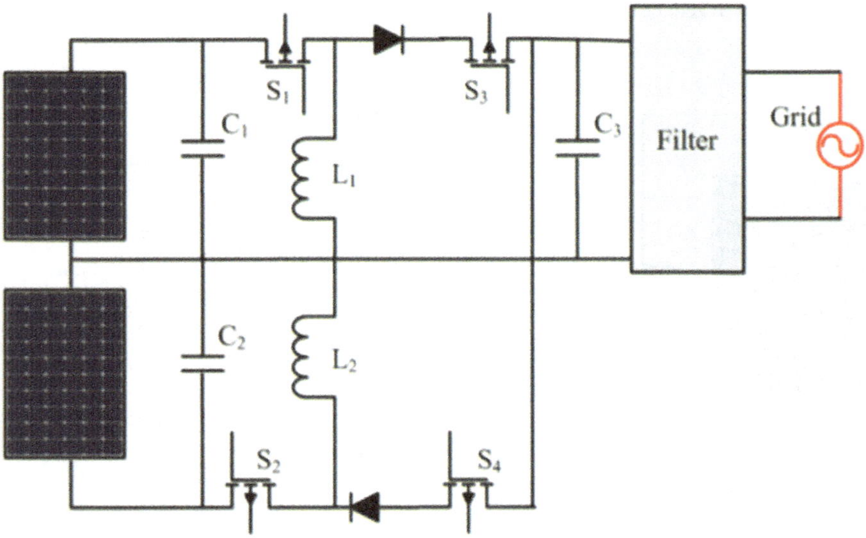

Fig. 31 Two-sourced, anti-parallel buck–boost inverter [42]

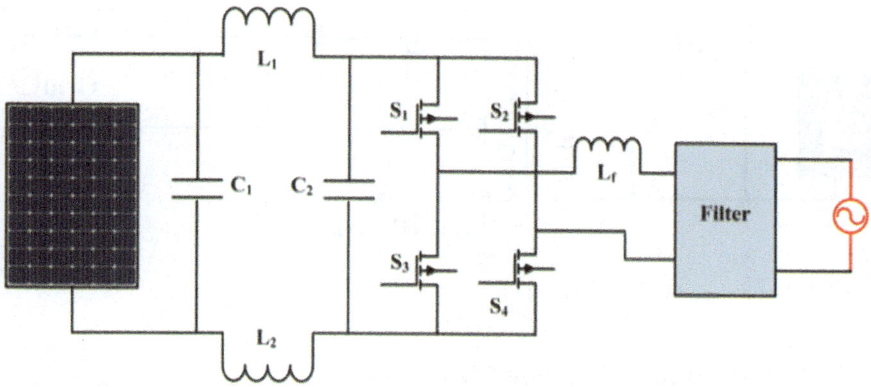

Fig. 32 Single phase Z-source inverter [44]

From the comparative analysis presented in Table 6 of various single-stage inverter topologies, the authors concluded that the structures presented in Table 8 were more efficient compared to other topologies.

Table 8 Analysis on single-stage inverter topologies

Figure No. and Reference	V_{in} (V)	V_{out} (V)	Power rating (W)	Frequency (kHz)	Switching time (ms)	Maximum efficiency (%)
Figure 26 and [66]	16.8	110	200	10	0.1	70
Figure 27 and [38]	20	230	170	10	0.1	85
Figure 28 and [107]	200–350	230	1000	–	–	93.6
Figure 29 and [39]	60–75	120	180	15	0.15	–
Figure 30 and [40]	50	120	200	–	–	75
Figure 31 and [41]	40–50	90–110	500	9.2	0.092	80–90
Figure 32 and [42]	50–250	120	950	5	0.05	–
Figure 33 and [44]	150	120	200	20	0.2	94–95

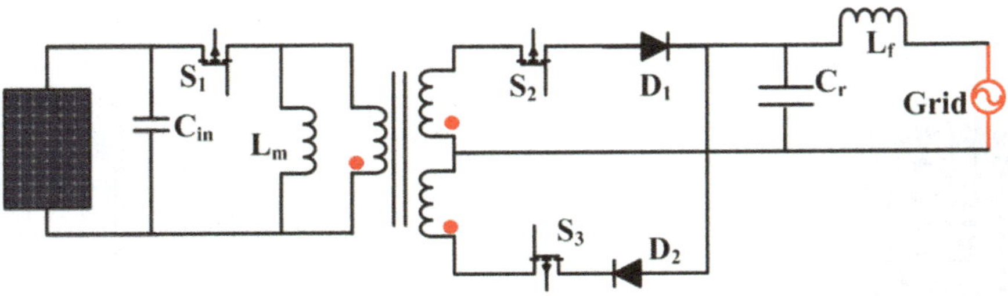

Fig. 33 Single stage fly back inverter [48]

3.7 *Grid-Connected Isolated Inverters*

Micro-inverters have been segregated into single-stage and multistage topologies. Single-stage flyback converters with lower semiconductor equipment [47] are widely used. The different topologies under of this type are explained below.

3.7.1 Single Fly Back Topologies

Figure 33 shows a fly back inverter with a center-tapped transformer [48]. The benefit of fly back topology is its capacity to combine the energy storage device with the transformer [48]. The grouping of these 2 elements in a fly back topology eradicates the DC current sensor and minimizes cost.

3.7.2 Fly Back Converter with Power Decoupling Circuit

The control fly back type single-stage micro-inverter employing the DCM mode has been proposed by the authors [49] and is illustrated in Fig. 34. Maximum efficiency attained by this topology is only 70% due the loss of power on the MOSFET.

3.7.3 Three-Port Fly Back Inverter with Coupling Circuit

Figures 35 and 36 illustrate the three-port power coupling and decoupling circuits. The power decoupling switch is present at the third port in Fig. 35. Due to the presence of the high voltage, coupling capacitor is reduced [50].

3.7.4 Soft-Switching Single-Stage Fly Back Converter

Figure 37 displays the soft-switching single fly back inverter [51]. From the circuit it can be observed that the bi-directional switches are placed on the secondary side of

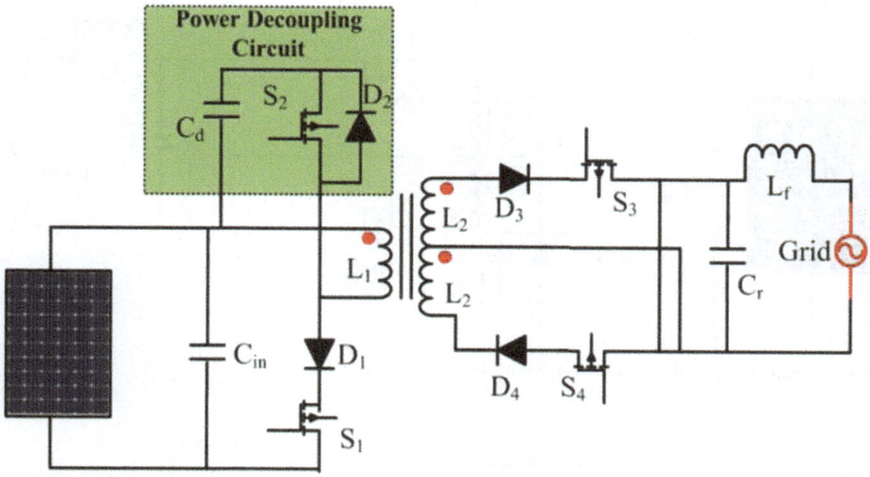

Fig. 34 Single stage fly back inverter with decoupling circuit [49]

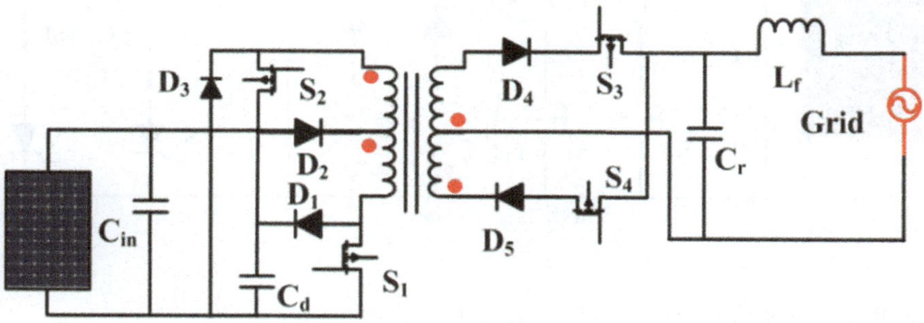

Fig. 35 Three-port flyback inverter with coupling circuit-1 [50]

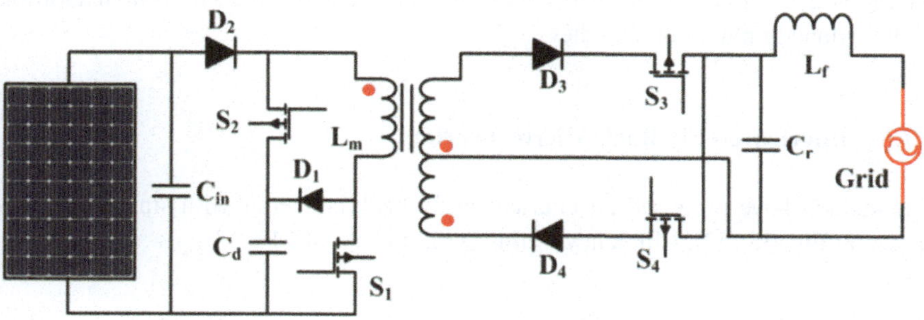

Fig. 36 Three-port fly back inverter with coupling circuit-2 [67]

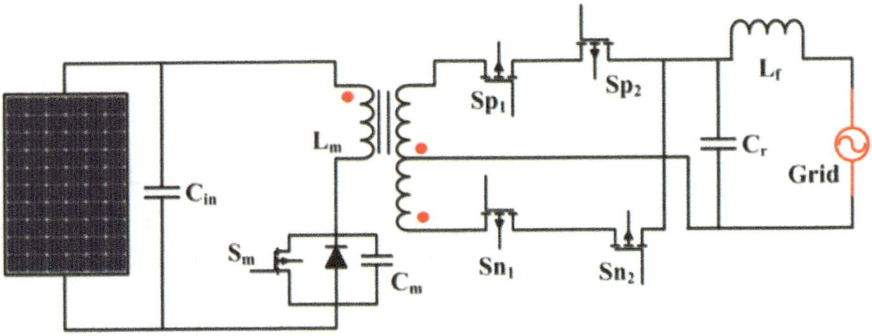

Fig. 37 Single stage fly back inverter with soft switching [51]

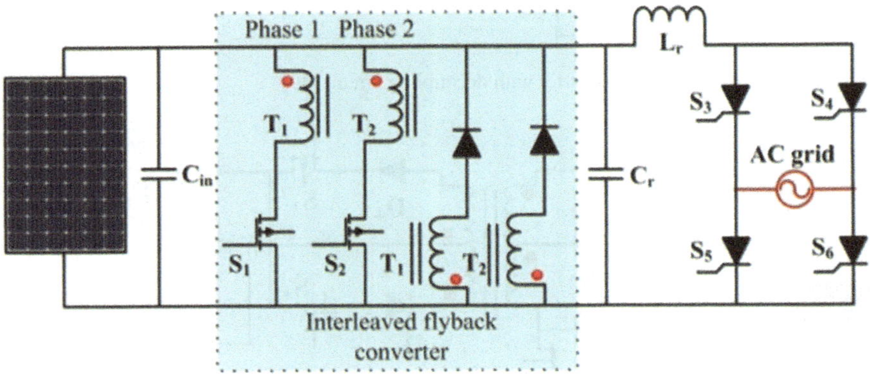

Fig. 38 Single-phase and two-phase interleaved fly back converter [52]

the topology. With this type of topology, efficiency can reach up to 95% in addition to the reduced number of switches.

3.7.5 Interleaved Fly Back Micro-Inverter

The authors have proposed an interleaved fly back converter with single phase as shown in Fig. 38 [52]. Efficiency of this system is about 94%.

3.7.6 Interleaved Converter with CMM Control (Flyback)

The topology presented in Fig. 39 shows the CMM control interleaved flyback converter. The resonant peak mode was affected when operated in the CMM mode.

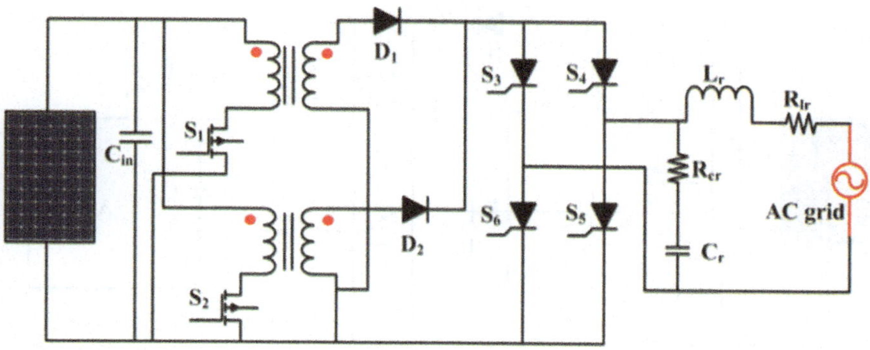

Fig. 39 CMM control interleaved flyback converter [68]

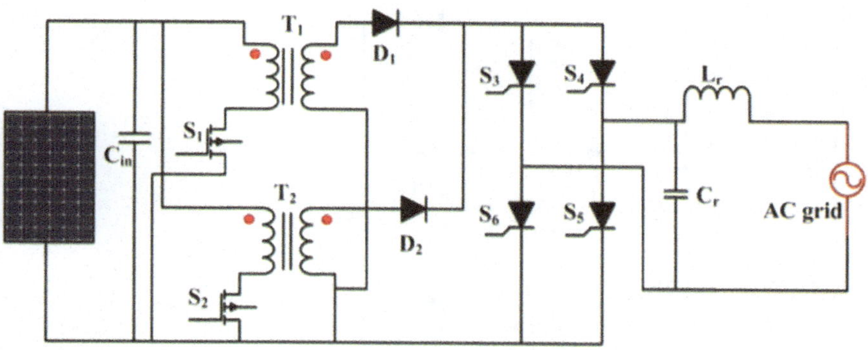

Fig. 40 DCM control interleaved flyback converter [53]

3.7.7 DCM Mode of the Interleaved Flyback Converter

Figure 40 illustrates the DCM mode of the interleaved flyback converter. Efficiency of this system can reach a maximum of up to 94% and THD is only 2.49%, as proposed by the authors [53].

3.7.8 Primary-Parallel Secondary Series Multi-core Inverter and Flyback Converter with Soft Switching

Another topology of micro-inverter, presented in Fig. 42, shows the primary-parallel, secondary series multicore-inverter proposed by the authors [54] where the stress is minimum. Figure 41 details the interleaved flyback micro-inverter with soft switching. The efficiencies of the topologies have been presented by authors to be about 95.1% [55, 56].

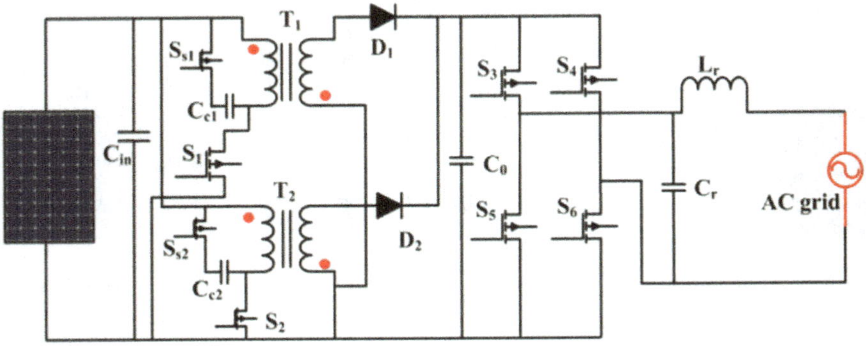

Fig. 41 Interleaved flyback inverter with soft switching [56]

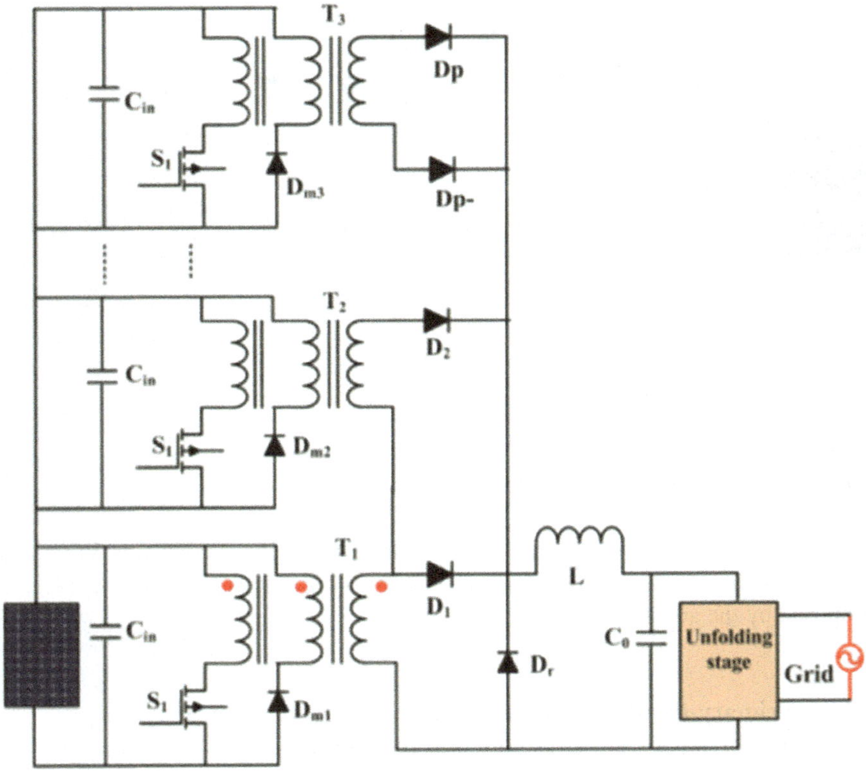

Fig. 42 Primary-parallel secondary series multi-core inverter [54]

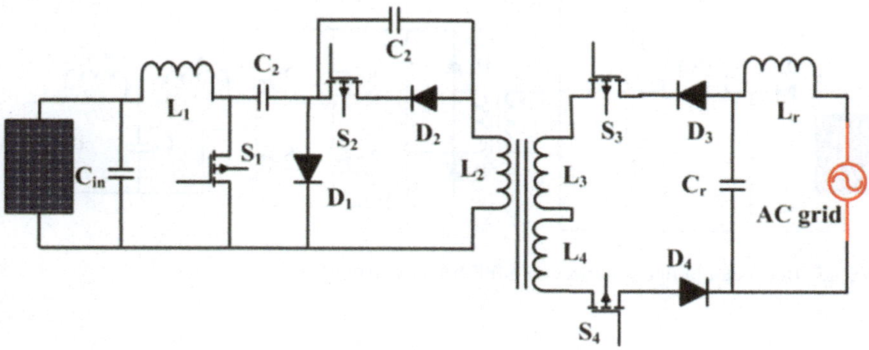

Fig. 43 Double-stage flyback inverter with soft switching [69]

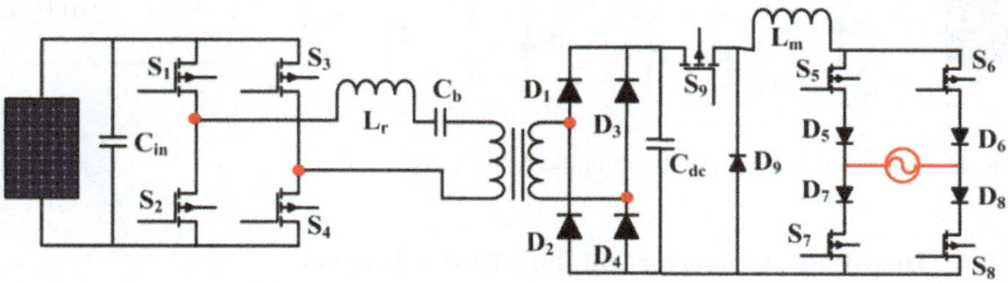

Fig. 44 Three-stage flyback inverter with soft switching [57]

3.8 Multistage Isolated Micro-Inverters

3.8.1 Soft-Switching Double-Stage Flyback Inverter

The topology presented in Fig. 43 shows the double-stage flyback micro-inverter with soft switching.

3.8.2 Soft-Switching Three-Stage Flyback Inverter

The three-stage flyback inverter with soft-switching technique is shown in Fig. 44 as proposed by the authors [57].

3.8.3 Full Bridge-Half Bridge Converter (Boost Mode)

Figure 45 shows the boost half bridge converter with full-bridge inverter proposed by the authors [58].

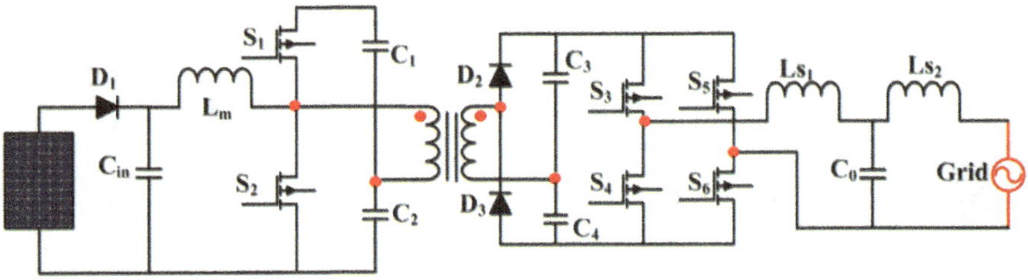

Fig. 45 Boost half bridge converter with full-bridge inverter [58]

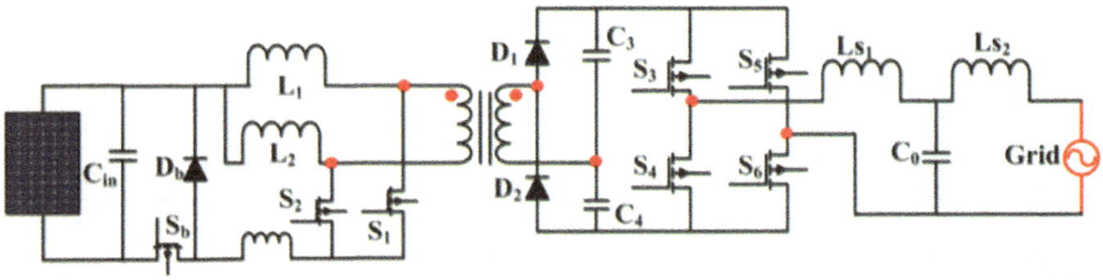

Fig. 46 Dual-boost converter with full-bridge inverter [59]

3.8.4 Dual-Boost Converter with Full-Bridge Inverter

Another new type of topology, called the dual-boost converter with full-bridge inverter, is shown in Fig. 46 [59].

3.8.5 Current-Fed Push-Pull Converter with Full-Bridge Inverter

Figure 47 shows the current-fed push-pull converter with full bride inverter [60]. This system has an efficiency of 97.5%.

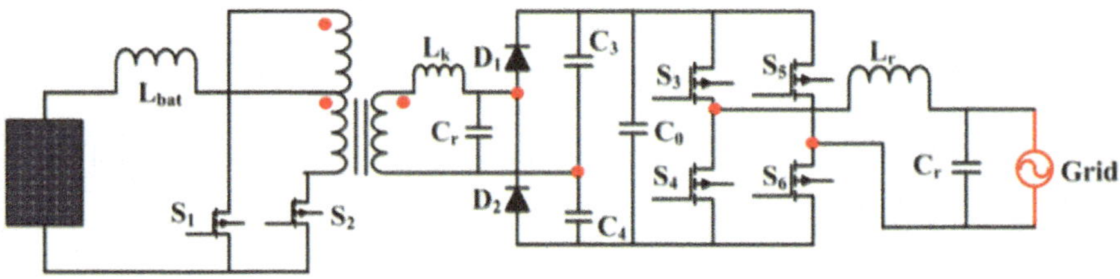

Fig. 47 Current-fed push-pull converter with full bridge inverter [60]

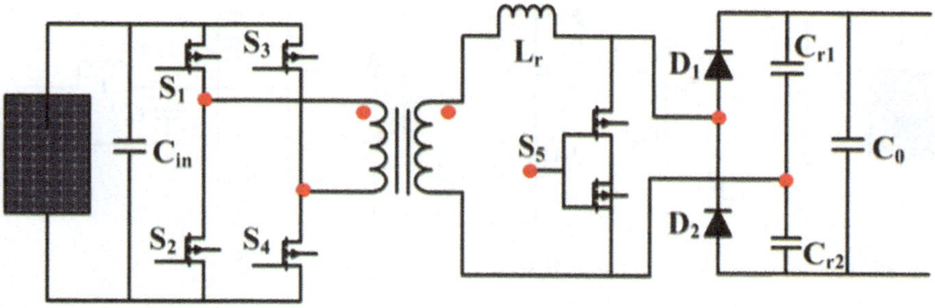

Fig. 48 Hybrid resonant DC-DC converter with soft switching [61]

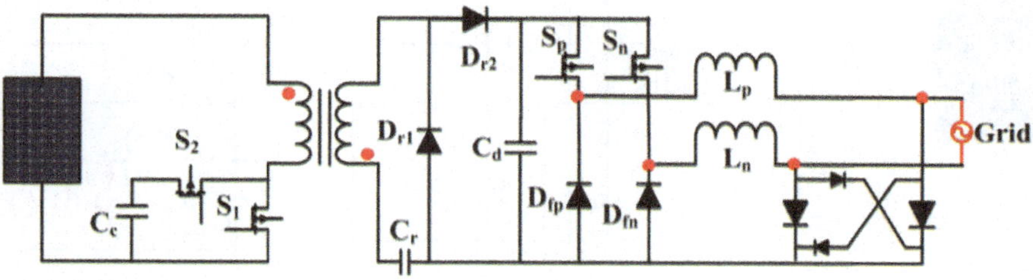

Fig. 49 Active clamp DC-DC converter with single switch modulated inverter [62]

3.8.6 Hybrid Resonant DC-DC Converter with Soft Switching

Another new converter topology used in PV applications is shown in Fig. 48. This hybrid resonant DC-DC converter has been proposed by the authors [61] with soft switching. This converter operates in three modes: (1) buck, (2) boost, and (3) series resonant modes. Overall efficiency of the system is about 97% at high power.

3.8.7 Active Clamp DC-DC Converter with Single Switch Modulated Inverter

Figure 49 shows the active DC-DC clamp converter with a single inverter modular switch. This inverter enables ZVS to be obtained by recycling the energy stored in the transformer's leakage induction, as described by the writers [62].

3.8.8 Flyback Converter with High Frequency AC (HFAC)-Linked Active Decoupling Circuit

The flyback inverter with HFAC-linked active decoupling circuit is shown in Fig. 50 [63]. Figure 51 also highlights the full-bridge resonant converter with three-phase converters.

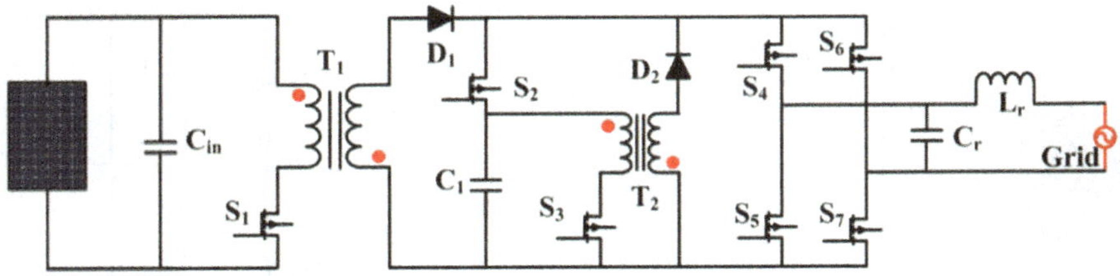

Fig. 50 Flyback converter with HFAC [63]

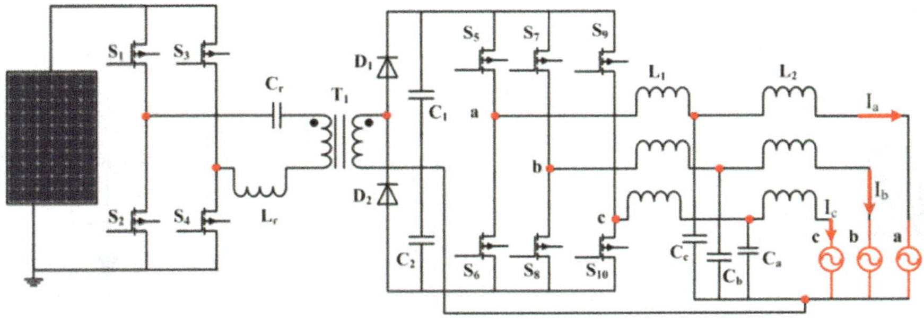

Fig. 51 Full-bridge resonant converter with three-phase inverter [70]

For a better understanding, various single-stage isolated PV inverters were compared with various parameters; results are presented in Table 9. Table 10 shows the comparative study of the multistage isolated PV micro-inverters.

4 Grid Standards and Guidelines

The amount of power generated by solar plants varies from 100 W to 1000 MW. Hence, PV plants could follow specific standards and regulations. These standards may vary across different countries. The grid standards and regulations are crucial factor which has a huge impact on performance and design of any PV plants.

There exist more famous institutions which framed grid standards that are followed by local grid companies in most of the countries. The institutions are namely IEC (International Electrotechnical Commission) in Switzerland, IEEE (Institute of Electrical and Electronic Engineers) in the US and DKE (German Commission for Electrical, Electronic and Information Technologies of DIN and VDE) in Germany. Grid connected PV plants mainly face issues on power factor maintained, total harmonic distortion, harmonic levels, amount of leakage current, fluctuations in voltage, current and frequency [64]. By noticing the importance of grid standards, the most widely used standards over harmonic level voltage, current and frequency fluctuations are tabulated in Table 11.

Table 9 Comparison of single stage isolated PV inverters

Figure No. and Reference	Power (W)	Frequency (kHz)	Mode of operation	Number of switches	Cost	THD (%)	Efficiency (%)	Life span
Figure 34 and [48]	300	9.6	DCM	3	low	–	89.1	S
Figure 35 and [49]	100	50	DCM	4	M	–	70.2	L
Figure 36 and Ref. [50]	100	50	DCM	4	M	<1.7	90.51	L
Figure 37 and [67]	100	50	Modified DCM	4	M	<1.9	90.12	L
Figure 38 and [51]	250	27–65	2-f and 1-f DCM	3	low	–	94.1	S
Figure 39 and [52]	200	100	DCM	6	M	<3.79	94.21	S
Figure 40 and [68]	200	170	CCM	6	M	<4	95.1	S
Figure 41 and [53]	200	170	BCM	6	M	<2.46	94.2	S
Figure 42 and [56]	250	300–400	DCM	8	M	–	95.1	S
Figure 43 and [54]	120	70	BCM and DCM	12	high	<5	92.43	L

*Where S—Short, L—Long, M—Medium
Note: L, M, H represents Low, Medium, and High respectively.

Table 10 Comparison of Multi- stage isolated PV micro-inverters

Figure No. and Reference	Power (W)	Frequency (kHz)	Mode of operation	Number of switches	Cost	THD (%)	Efficiency (%)	Life span
Figure 44 and [69]	500	20	DCM	4	M	<1.7	90.1	L
Figure 45 and [57]	150	25	–	9	H	<1.9	90	L
Figure 46 and [58]	210	10.8	–	6	H	–	94.1	L
Figure 47 and [59]	250	–	–	7	H	<3.79	94.1	S
Figure 52 and [70]	400	20–185	Modified DCM	10	H	<5	92	L

*Where S—Short, L—Long, M—Medium, H—High
Note: L, M, H represents Low, Medium, and High respectively.

Table 11 Widely used grid standards for effective operation PV plants

Reference	Name of standard	Power rating capacity		Harmonic level	DC current	Voltage variation		Frequency variation	
Ref. [108]	IEEE 1547	30 kw	Order	%	< 1% of rated current	Order	%	Order	%
			3–9	4.0		V < 50%	0.1 s	49 Hz < f < 49 Hz	0.2 s
			11–15	2.0		50% < V<88%	2 s		
			17–21	1.5		110% < V<120%	2 s		
			23–33	0.6		V > 120%	0.05 s		
			>35	0.3					
			Even harmonics	<25% of odd harmonics					
Ref. [108]	IEC 61727	10 kw	Order	%	< 0.5% of rated current	order	%	order	%
			3–9	4.0		V < 50%	0.16 s	59.3 Hz < f < 60.5 Hz	0.16 s
			11–15	2.0		50% < V<88%	2 s		
			17–21	1.5		110% < V<120%	1 s		
			23–33	0.6		V > 120%	0.16 s		
			>35	0.3					
			Even harmonics	<25% of odd harmonics					
Ref. [109]	VDE0126-1-1	–	Order	A	< 0.22 A	order	%	order	%
			3	3		V < 85%	0.2 s	47.5 Hz < f < 50.2 Hz	0.2 s
			5	1.5		V > 110%	0.22 s		
			7	1					
			9	0.7					
			11	0.33					
			Even harmonics	<1.5/h					

5 Recent Advancements–Challenges and Discussion

The developed new topologies attain higher efficiency and have less EMI effect. The effective design and development of converter topologies are not easy, the various parameters taken into consideration that are namely; reliability, conversion ratio, efficiency, thermal stress, conduction loss etc. [65]. Further, the selection of inductor and capacitors are challenging because they act as filters. The switching frequency is another major factor which decides the size of the converter. Few other parameters which need to be taken into account are noise and dissipation of heat.

To provide a quick look, performance parameters and elements involved in the design of various converter topologies are presented in Tables 3 and 4 for the benefit of reader. In addition to the traditional converter topologies, the authors have analyzed more recent converter topologies which can be used for the solar PV application. After performing a critical analysis, a comparison is made by considering various important parameters such as output voltage, number of switching devices, features and other elements used in the development of converter topologies. This comparative study is presented in Table 12.

6 Direction for Future Research

The integration of renewable energy resources into a grid connected system can be implemented in single and two stage conversion. Moreover, the traditional DC-DC converters are not preferred due to limitation of high gain. Further, the two-stage conversion system requires two conventional DC-DC converters and one DC-AC inverter. These types of topologies are of huge size, significant expensive and less efficient. To overcome these limitations the concept of multi-port converters can develop.

The understanding of converter and inverter topologies are mandatory for integrating renewable energy resources with the grid connected systems and following the grid standards to maintain minimum harmonic levels are essential. At present, less efficient PV modules are available in the market, to guarantee high efficiency research on materials for the manufacture of new PV modules are required. Furthermore, overall efficiency of grid connected system will improve which reduces the power generation cost. The novel configurations of on grid inverters are in a developing stage since a decade ago. In addition to the conventional converter topologies, it is necessary to focus and develop high gain multi-port converter topologies for effective usage. Solar PV systems such as tri-switching state non-isolated high gain boost converter, high voltage gain SEPIC converter, transformer and switched capacitor-based extension for boost converter, inverting modified CUK converter configurations, modular high gain converters. The other inverter topologies which needs more attention for effective grid integration are multilevel transformer less inverters, evolution of power storage converters which has capability of LVRT, future power

Table 12 Comparative analysis of different types of converter topologies

Type of converter	Output voltage equation	Number of switching devices		Number of capacitors	Number of inductors	Number of transformers	Features				Reference
		Diodes	Switches				Soft-switching	Battery charging	Isolation	Source utility	
Multiport Bidirectional Converter	$V_0 = \frac{n_3}{n_1} V_{S1} + \frac{n_2}{n_1} V_{S2}$		6	7	0	1 three-winding	√	–	√	√	Refs. [110, 111]
Multi-input converter as a BLDC drive	$V_0 = \frac{V_1 + d_4 V_3}{1 - d_1 - d_4}$ $V_0 = \frac{V_2 - d_3 V_3}{1 - d_2 - d_3}$	2	4	1	2	0		√		√	Ref. [112]
Non-isolated Multi-input Single-output DC/DC Buck/Boost-Boost converter	$V_{0ut} = \frac{E_2}{1-d_2} + \frac{d_1 d_2}{(1-d_1)(1-d_2)} E_1$	2	2	2	2	0				√	Ref. [112]
Buck/Boost Multi-input converter	$V_0 = \frac{d_1}{N_1(1 - d_1 - d_2)} V_1 + \frac{d_2}{N_2(1 - d_1 - d_2)} V_2$	3	2	1	0	1 three-winding	–	–	√	–	Refs. [112–114]
Full-bridge Boost Converter Based the Distribution Transformer	$V_0 = \frac{n}{2(1-d_1)} V_1$ $V_0 = \frac{n}{2(1-d_2)} V_2$	8	8	2	2	1 eight-winding			√		Refs. [115, 116]
Double input Full-bridge converter	$V_0 = \frac{n_3}{n_1} V_1 + \frac{n_2}{n_1} V_2$	12	8	1	2	1 three-winding			√	√	Refs. [117, 118]

(continued)

Table 12 (continued)

Type of converter	Output voltage equation	Number of switching devices		Number of capacitors	Number of inductors	Number of transformers	Features				Reference
		Diodes	Switches				Soft-switching	Battery charging	Isolation	Source utility	
Direct charge converter	$V_2 = \frac{d_1}{d_1+d_2} V_1$ $V_0 = 2n\, d_1\, V_2$	0	4	5	0	1 two-winding		√	√	√	Refs. [119, 120]
Double input Buck/Buck-Boost converter	$V_0 = \frac{d_1}{1-d_2} V_1 + \frac{d_2}{1-d_2} V_2$	2	2	1	1	0	√			√	Refs. [121–124]
Series or parallel of two DC/DC converter	$V_0 = \frac{V_1}{1-d_1} + \frac{V_2}{1-d_2}$ $V_0 = \frac{V_1}{1-d_2} + \frac{V_2}{1-d_4}$	2 0	2 4	2	2	0				√	Refs. [125–130]
Non-isolated Multi-input Multi-output DC/DC Boost converter	–	4	4	2	1	0		√		√	Refs. [131]
Combined DC Link And Magnetic Couple Converter	$D = \frac{V_{DC}-V_{SC}}{V_{SC}}$ $V_{DC} = 2V_2$ $V_{Load} = n\, V_{DC}$	0	6	5	2	1 two-winding	√	√	√	√	Refs. [108, 132]

(continued)

Table 12 (continued)

Type of converter	Output voltage equation	Number of switching devices		Number of capacitors	Number of inductors	Number of transformers	Features				Reference
		Diodes	Switches				Soft-switching	Battery charging	Isolation	Source utility	
Full-Bridge Three-Port Converters With Wide Input Voltage Range	$V_1 = \frac{d_2}{d_1} V_1$ $V_0 = 2n\, d_3\, V_2$	4	4	1	1	1 two-winding		√	√		Refs. [132]

conditioner quasi Z source networks. The increasing focus in this field enhances the consistency, power density and overall performance of power electronic converters.

The growth in the development of thin film PV modules and wide bandgap gadges will presumably direct the research in new ways and hence transforming the sense of the best across the widespread converter topologies.

7 Summary/Conclusion

This article presented a complete topologies outline of converters/inverters used in standalone and grid-connected PV systems based on their control procedures. Focusing on converters/inverters used in PV systems, a comparison was made based on the general characteristics of converter topologies. The various converter topologies used to track maximum power were analyzed and presented features and remarks about each MPPT method are discussed. To further enhance knowledge of these topologies, comparative tables were presented, considering parameters that will influence the performance of any converter topology, such as losses, efficiency, cost, etc. The authors also focused on reviewing the most widely used inverter topologies for application in solar PV systems. Numerous inverter topologies were presented thorough analysis of their various factors, namely power rating, efficiency, switching time, number of switches THD, and lifespan of the inverters. This should help the reader to choose a suitable converter/inverter topology according to their applications. This article can serve as a "one-stop shopping" solution for all researchers and engineers working on converter/inverter topologies to work with a particular converter/inverter for their specific application. In addition, the authors expect that, this research article will prove as a benchmark for practitioners, and researchers in the area of converter/inverter topologies.

References

1. WEO, World energy outlook. International Energy Agency (IEA) (2018). https://www.power-technology.com/features/world-energy-outlook-2018-fossil-fuels/
2. T. Sudhakar Babu, K. Sangeetha, N. Rajasekar, Voltage band based improved particle swarm optimization technique for maximum power point tracking in solar photovoltaic system. J. Renew. Sustain. Energy **8**, 013106 (2016)
3. T.S. Babu, N. Rajasekar, K. Sangeetha, Modified particle swarm optimization technique based maximum power point tracking for uniform and under partial shading condition. Appl. Soft Comput. **34**, 613–624 (2015)
4. J.P. Ram, T. Sudhakar Babu, N. Rajasekar, A comprehensive review on solar PV maximum power point tracking techniques. Renew. Sustain. Energy Rev. 67, 826–847 (2017)
5. Z. Yang, P.C. Sen, A novel switch-mode DC-to-AC inverter with nonlinear robust control. IEEE Trans. Ind. Electron. **45**(4), 602–608 (1998)

6. J.K. Kim, G.W. Moon, Derivation, analysis, and comparison of non-isolated single-switch high step-up converters with low voltage stress. IEEE Trans. Power Electron. **30**(3), 1336–1344 (2014)
7. R. Teodorescu, F. Blaabjerg, M. Liserre, P.C. Loh, Proportional-resonant controllers and filters for grid-connected voltage-source converters. IEEE Proc. Electric Power Appli. **153**(5), 750–762 (2006)
8. W. Baochao, M. Sechilariu, F. Locment, Intelligent DC microgrid with smart grid communications: control strategy consideration and design. IEEE Trans. Smart Grid **3**(4), 2148–2156 (2012)
9. O. Veneri, C. Capasso, D. Iannuzzi, Experimental evaluation of DC charging architecture for fully-electrified low-power two-wheeler. Appl. Energy **162**, 1428–1438 (2016)
10. Y. Sun, C. Zhong, X. Hou, J. Yang, H. Han, J.M. Guerrero, Distributed cooperative synchronization strategy for multi-bus microgrids. Int. J. Electr. Power Energy Syst. **86**, 18–28 (2017)
11. W. Li, W. Li, X. He, D. Xu, B. Wu, General derivation law of non-isolated high-step-up interleaved converters with built-in transformer. IEEE Trans. Ind. Electron. **59**(3), 650–1661 (2011)
12. B. Nayak, A. Mohapatra, K.B. Mohanty, Selection criteria of dc-dc converter and control variable for MPPT of PV system utilized in heating and cooking applications. Cogent Eng. **4**(1) 1363357 (2017)
13. F. Blaabjerg, Z. Chen, S.B. Kjaer, Power electronics as efficient interface of renewable energy sources, in *The 4th International Power Electronics and Motion Control Conference, 2004* (IPEMC, 2004), vol. 3, 1731–1739
14. T. Esram, P.L. Chapman, Comparison of photovoltaic array maximum power point tracking techniques. IEEE Trans. Energy Conver. **22**(2), 439–449 (2007)
15. S. Wang, Design and hardware implementation of a soft-switched converter for fuel cell applications (2007)
16. O. Krykunov, Comparison of the dc/dc-converters for fuel cell applications, Int. J. Electri. Comput. Syst. Eng. 1(1), 71–79 (2007)
17. M.K. Kazimierczuk, D. Czarkowski. *Resonant Power Converters* (John Wiley & Sons, 2012)
18. S. Bansala, L.M. Saini, Analysis and comparison of various soft switching topologies for PSFB DC-DC converter with additional auxiliary circuits. Int. J. Electr. Eng. **5**(2), 1255–1268 (2014)
19. M.T. Outeiro, G. Buja, D. Czarkowski, Resonant power converters: An overview with multiple elements in the resonant tank network. IEEE Ind. Electron. Mag. **10**(2), 21–45 (2016)
20. A. Amir, A. Amir, H.S. Che, A. ElKhateb, N. Abd Rahim, Comparative analysis of high voltage gain DC-DC converter topologies for photovoltaic systems. Renew. Energy (2018)
21. R.R. Gopi, S. Sreejith, Converter topologies in photovoltaic applications–A review. Renew. Sustain. Energy Rev. 94, 1–14 (2018)
22. M.F. Rahman, L. Zhong, A new, transformerless, photovoltaic array to utility grid interconnection, in *Proceedings of Second International Conference on Power Electronics and Drive Systems,* vol. 1 (1997), pp. 139–143
23. R. Dogga, M.K. Pathak, Recent trends in solar PV inverter topologies. Solar Energy **183,** 57–73 (2019)
24. K. Zeb, W. Uddin, M.A. Khan, Z. Ali, M.U. Ali, N. Christofides, H.J. Kim, A comprehensive review on inverter topologies and control strategies for grid-connected photovoltaic system. Renew. Sustain. Energy Rev. **94**, 1120–1141 (2018)
25. M. Veerachary, T. Senjyu, K. Uezato, Neural-network-based maximum-power-point tracking of coupled-inductor interleaved-boost-converter-supplied pv system using fuzzy controller. IEEE Trans. Ind. Electron. **4**, 749–758 (2003)
26. M.G. Villalva, J.R. Gazoli, E. Ruppert Filho, Comprehensive approach to modeling and simulation of photovoltaic arrays. IEEE Trans. Power Electron. **24**(5), 1198–1208 (2009)
27. N. Vázquez, J. Almazán, J. Álvarez, C. Aguilar, J. Arau, Analysis and experimental study of the buck, boost and buck-boost inverters, in *30th Annual IEEE Power Electronics Specialists Conference. Record (Cat. No. 99CH36321)*, vol. 2, 801–806 (1999)

28. A. Cabrera-Tobar, E. Bullich-Massagué, M. Aragüés-Peñalba, O. Gomis-Bellmunt, Topologies for large scale photovoltaic power plants. Renew. Sustain. Energy Rev. 59, 309–319 (2016)
29. M.M. Hasan, A. Abu-Siada, S.M. Islam, S.M. Muyeen, A novel concept for three-phase cascaded multilevel inverter topologies. Energies **11**(2), 268 (2018)
30. A.M. Noman, A. Al-Shamma'a, K.E. Addoweesh, A.A. Alabduljabbar, A.I. Alolah, Cascaded multilevel inverter topology based on cascaded H-bridge multilevel inverter. Energies **11**(4), 895 (2018)
31. G. Sarowar, M.A. Choudhury, M.A. Hoque, A novel control scheme for buck-boost DC to AC converter for variable frequency applications. Proc. Soc. Behav. Sci. **195**, 2511–2519 (2015)
32. Y.H. Kim, J.G. Kim, Y.H. Ji, C.Y. Won, Y.C. Jung, Photovoltaic parallel resonant dc-link soft switching inverter using hysteresis current control, in *2010 Twenty-Fifth Annual IEEE Applied Power Electronics Conference and Exposition (APEC)*, 2275–2280 (2010)
33. J.M. Shen, H.L. Jou, J.C. Wu, Novel transformerless grid-connected power converter with negative grounding for photovoltaic generation system. IEEE Trans. Power Electron. **27**(4), 1818–1829 (2011)
34. J.L. Duran-Gomez, E. Garcia-Cervantes, D.R. Lopez-Flores, P.N. Enjeti, L. Palma, Analysis and evaluation of a series-combined connected boost and buck-boost dc-dc converter for photovoltaic application, in *Twenty-First Annual IEEE Applied Power Electronics Conference and Exposition, 2006 (APEC'06)* (2006), p. 7
35. R. Mallwitz, Circuit apparatus for transformerless conversion of an electric direct voltage into an alternating voltage, U.S. Patent No. 7,616,467 (2009)
36. S.V. Araujo, P. Zacharias, B. Sahan, Novel grid-connected non-isolated converters for photovoltaic systems with grounded generator, in *2008 IEEE Power Electronics Specialists Conference* (2008), pp. 58–65
37. S. Schoenbauer, F.R. Martin-Lopez, Single-inductor buck-boost converter with positive and negative outputs, U.S. Patent Application 12/228, 310 (2010)
38. L.G. Junior, M.A.G. De Brito, L.P. Sampaio, C.A. Canesin, Single stage converters for low power stand-alone and grid-connected PV systems, in *2011 IEEE International Symposium on Industrial Electronics (ISIE)* (2011), pp. 1112–1117
39. T.J. Liang, J.L. Shyu, J.F. Chen, A novel DC/AC boost inverter, in *IECEC'02. 2002 37th Intersociety Energy Conversion Engineering Conference* (2002), pp. 629–634
40. R.O. Caceres, I. Barbi, A boost DC-AC converter: analysis, design, and experimentation. IEEE Trans. Power Electron. **14**(1), 134–141 (1999)
41. M. Kusakawa, H. Nagayoshi, K. Kamisako, K. Kurokawa, Further improvement of a transformerless, voltage-boosting inverter for ac modules. Solar Energy Mater. Solar Cells **67**(1–4), 379–387 (2001)
42. N. Kasa, T. Iida, H. Iwamoto, An inverter using buck-boost type chopper circuits for popular small-scale photovoltaic power system, in *Proceedings. 25th Annual Conference of the IEEE Industrial Electronics Society (Cat. No. 99CH37029), IECON'99 Conference* vol. 1 (1999), pp. 185–190
43. H. Patel, V. Agarwal, A single-stage single-phase transformer-less doubly grounded grid-connected PV interface. IEEE Trans. Energy Convers. **24**(1), 93–101 (2009)
44. Y. Tang, S. Xie, C. Zhang, Single-phase Z-source inverter. IEEE Trans. Power Electron. **26**(12), 3869–3873 (2010)
45. D. Cao, S. Jiang, X. Yu, F.Z. Peng, Low-cost semi-Z-source inverter for single-phase photovoltaic systems. IEEE Trans. Power Electron. **26**(12), 3514–3523 (2011)
46. D. Li, P.C. Loh, M. Zhu, F. Gao, F. Blaabjerg, Generalized multicell switched-inductor and switched-capacitor Z-source inverters. IEEE Trans. Power Electron. **28**(2), 837–848 (2012)
47. S.B. Kjaer, J.K. Pedersen, F. Blaabjerg, A review of single-phase grid-connected inverters for photovoltaic modules. IEEE Trans. Ind. Appl. **41**(5), 1292–1306 (2005)
48. N. Kasa, T. Iida, L. Chen, Flyback inverter controlled by sensorless current MPPT for photovoltaic power system. IEEE Trans. Ind. Electron. **52**(4), 1145–1152 (2005)

49. S.B. Kjaer, F. Blaabjerg, Design optimization of a single phase inverter for photovoltaic applications, in *IEEE 34th Annual Conference on Power Electronics Specialist, 2003 (PESC'03)*, vol. 3 (2003), pp. 1183–1190
50. H. Hu, S. Harb, X. Fang, D. Zhang, Q. Zhang, Z.J. Shen, I. Batarseh, A three-port flyback for PV microinverter applications with power pulsation decoupling capability. IEEE Trans. Power Electron. **27**(9), 3953–3964 (2012)
51. N. Sukesh, M. Pahlevaninezhad, P.K. Jain, Analysis and implementation of a single-stage flyback PV microinverter with soft switching. IEEE Trans. Ind. Electron. **61**(4), 1819–1833 (2013)
52. Z. Zhang, X.F. He, Y.F. Liu, An optimal control method for photovoltaic grid-tied-interleaved flyback microinverters to achieve high efficiency in wide load range. IEEE Trans. Power Electron. **28**(11), 5074–5087 (2013)
53. Y. Li, R. Oruganti, A low cost flyback CCM inverter for AC module application. IEEE Trans. Power Electron. **27**(3), 1295–1303 (2011)
54. D. Meneses, O. Garcia, P. Alou, J.A. Oliver, J.A. Cobos, Grid-connected forward microinverter with primary-parallel secondary-series. IEEE Trans. Power Electron. **30**(9), 4819–4830 (2014)
55. Y.H. Kim, Y.H. Ji, J.G. Kim, Y.C. Jung, C.Y. Won, A new control strategy for improving weighted efficiency in photovoltaic AC module-type interleaved flyback inverters. IEEE Trans. Power Electron. **28**(6), 2688–2699 (2012)
56. M. Gao, M. Chen, C. Zhang, Z. Qian, Analysis and implementation of an improved flyback inverter for photovoltaic ac module applications. IEEE Trans. Power Electron. **29**(7), 3428–3444 (2013)
57. C. Rodriguez, G.A.J. Amaratunga, Long-lifetime power inverter for photovoltaic AC modules. IEEE Trans. Ind. Electron. **55**(7), 2593–2601 (2008)
58. S. Jiang, D. Cao, Y. Li, F.Z. Peng, Grid-connected boost-half-bridge photovoltaic microinverter system using repetitive current control and maximum power point tracking. IEEE Trans. Power Electron. **27**(11), 4711–4722 (2012)
59. H.J. Chiu, Y.K. Lo, C.Y. Yang, S.J. Cheng, C.M. Huang, C.C. Chuang, M.C. Kuo, Y.M. Huang, Y.B. Jean, Y.C. Huang, A module-integrated isolated solar microinverter. IEEE Trans. Ind. Electron. **60**(2), 781–788 (2013)
60. Y.H. Kim, S.C. Shin, J.H. Lee, Y.C. Jung, C.Y. Won, Soft-switching current-fed push–pull converter for 250-W AC module applications. IEEE Trans. Power Electron. **29**(2), 863–872 (2013)
61. T. LaBella, J.S. Lai, A hybrid resonant converter utilizing a bidirectional GaN AC switch for high-efficiency PV applications. IEEE Trans. Ind. Appl. **50**(5), 3468–3475 (2014)
62. W.J. Cha, Y.W. Cho, J.M. Kwon, B.H. Kwon, Highly efficient microinverter with soft-switching step-up converter and single-switch-modulation inverter. IEEE Trans. Industr. Electron. **62**(6), 3516–3523 (2014)
63. L. Korosec, T. Konjedic, M. Truntic, M. Rodic, M. Milanovic, PDM flyback PV microinverter with HFAC-link and active decoupling circuit. Electron. Lett. **51**(6), 516–517 (2015)
64. D. Meneses, F. Blaabjerg, Ó. García, J.A. Cobos, Review and comparison of step-up transformerless topologies for photovoltaic AC-module application. IEEE Trans. Power Electron. **28**, 2649–2663 (2013)
65. IEEE, IEEE Application Guide for IEEE Std 1547(TM), IEEE Standard for Interconnecting Distributed Resources with Electric Power Systems; IEEE Std 15472–2008; (IEEE, Piscataway, NJ, USA, 2009); pp. 1–217
66. B.S. Prasad, S. Jain, V. Agarwal, Universal single-stage grid-connected inverter. IEEE Trans. Energy Convers. **23**(1), 128–137 (2008)
67. H. Hu, S. Harb, N.H. Kutkut, Z.J. Shen, I. Batarseh, A single-stage microinverter without using eletrolytic capacitors. IEEE Trans. Power Electron. **28**(6), 2677–2687 (2012)
68. F.F. Edwin, W. Xiao, V. Khadkikar, Dynamic modeling and control of interleaved flyback module-integrated converter for pv power applications. IEEE Trans. Ind. Electron. **61**(3), 1377–1388 (2013)

69. T. Shimizu, K. Wada, N. Nakamura, Flyback-type single-phase utility interactive inverter with power pulsation decoupling on the dc input for an ac photovoltaic module system. IEEE Trans. Power Electron. **21**(5), 1264–1272 (2006)
70. L. Chen, A. Amirahmadi, Q. Zhang, N. Kutkut, I. Batarseh, Design and implementation of three-phase two-stage grid-connected module integrated converter. IEEE Trans. Power Electron. **29**(8), 3881–3892 (2013)
71. J. Lopez-Seguel, S.I. Seleme, P. Donoso-Garcia, L.F. Morais, P. Cortizo, M.S. Mendes, Comparison of MPPT approaches in autonomous photovoltaic energy supply system using DSP in *2010 IEEE International Conference on Industrial Technology* (2010), pp. 1149–1154
72. A. Ayachit, D.K. Saini, M.K. Kazimierczuk, Two-phase buck DC-AC converter as dynamic power supply for amplitude-modulated Class-DE Power Amplifier, in *2015 IEEE 58th International Midwest Symposium on Circuits and Systems (MWSCAS)*, (2015), pp. 1–4
73. D.W. Hart, *Power Electronics* (Tata McGraw-Hill Education, 2011)
74. S. Mageshwari, S. Kanagalakshmi, H. Rao, Design and implementation of buck-boost converter for residential PV application. Int. J. Electr. Eng. Inform. **9**(4), 834–849 (2017)
75. J.R. Maglin, R. Ramesh, Design and analysis of SEPIC converter based MPPT for solar PV module with CPWM. J. Electr. Eng. Technol. **9**(4), 1269–1276 (2014)
76. M. Veerachary, Power tracking for nonlinear PV sources with coupled inductor SEPIC converter. IEEE Trans. Aerosp. Electr. Syst. **41**(3), 1019–1029 (2005)
77. C. Vel, T. Venkatesan, Analysis of non-isolated multi-port single ended primary inductor converter or standalone applications. Energies **11**(3), 539 (2018)
78. R.D. Middlebrook, S. Cuk, A general unified approach to modelling switching-converter power stages, in *1976 IEEE Power Electronics Specialists Conference* (1976), pp. 8–34
79. C. Ankiha, Gopalaiah, Single phase cuk rectifier to get positive output voltage and reduced total harmonic distortion. Int. J. Adv. Res. Electron. Commun. Eng. (IJARECE) **5**, 1602–1608 (2016)
80. E. Abdeen, M.A. Gaafar, M. Orabi, E.M. Ahmed, A. El Aroudi, Multi-Input Ćuk-derived buck-boost voltage source inverter for photovoltaic systems in microgrid applications. Energies **12** (10), 2007
81. R. Kumar, B. Singh, BLDC motor-driven solar PV array-fed water pumping system employing zeta converter. IEEE Trans. Ind. Appl. **52**(3), 2315–2322 (2016)
82. B. Singh, R. Kumar, Solar photovoltaic array fed water pump driven by brushless DC motor using Landsman converter. IET Renew. Power Gener **10**(4), 474–484 (2016)
83. J.K. Kim, S.W. Choi, G.W. Moon, Zero-voltage switching post regulation scheme for multi-output forward converter with synchronous switches. IEEE Trans. Ind. Electron. **58**(6), 2378–2386 (2010)
84. M.E. Basoglu, B. Cakir, Hardware based comparison of buck-boost converter topologies in MPPT systems, in *2015 9th International Conference on Electrical and Electronics Engineering (ELECO)*, 1109–1112 (2015)
85. S.J. Chiang, H.J. Shieh, M.C. Chen, Modeling and control of PV charger system with SEPIC converter. IEEE Trans. Ind. Electron. **56**(11), 4344–4353 (2008)
86. T. Kok Soon, S. Mekhilef, A. Safari, Simple and low cost incremental conductance maximum power point tracking using buck-boost converter. J. Renew. Sustain. Energy **5**(2), 023106 (2013)
87. N.H. Baharudin, T.M. Mansur, F.A. Hamid, R. Ali, M.I. Misrun, Topologies of DC-DC converter in solar PV applications. Indon. J. Electr. Eng. Comput. Sci. **8**(2), 368–374 (2017)
88. P.F. de Melo, R. Gules, E.F.R. Romaneli, R.C. Annunziato, A modified SEPIC converter for high-power-factor rectifier and universal input voltage applications. IEEE Trans. Power Electron **25**(2), 310–321 (2010)
89. C. Hua, C. Shen, Study of maximum power tracking techniques and control of DC/DC converters for photovoltaic power system, in *PESC 98 Record. 29th Annual IEEE Power Electronics Specialists Conference (Cat. No. 98CH36196)*, vol. 1 (1998), pp. 86–93
90. R. Gules, W.M. Dos Santos, F.A. Dos Reis, E.F.R. Romaneli, A.A. Badin, A modified SEPIC converter with high static gain for renewable applications. IEEE Trans. Power Electron. **29**(11), 5860–5871 (2013)

91. S. Bhattacharjee, B.J. Saharia, A comparative study on converter topologies for maximum power point tracking application in photovoltaic generation. J. Renew. Sustain. Energy **6**(5), 053140 (2014)
92. J.M. Enrique, E. Duran, M. Sidrach-de-Cardona, J.M. Andujar, Theoretical assessment of the maximum power point tracking efficiency of photovoltaic facilities with different converter topologies. Solar Energy **81**(1), 31–38 (2007)
93. E. Mamarelis, G. Petrone, G. Spagnuolo, Design of a sliding-mode-controlled SEPIC for PV MPPT applications. IEEE Trans. Ind. Electron. **61**(7), 3387–3398 (2013)
94. S. Poshtkouhi, V. Palaniappan, M. Fard, O. Trescases, A general approach for quantifying the benefit of distributed power electronics for fine grained MPPT in photovoltaic applications using 3-D modeling. IEEE Trans. Power Electron. **27**(1), 4656–4666 (2011)
95. R. Rajesh, M.C. Mabel, A comprehensive review of photovoltaic systems **51**, 231–248 (2015)
96. Q. Zhao, F.C. Lee, High-efficiency, high step-up DC-DC converters. IEEE Trans. Power Electron. **18**(1), 65–73 (2003)
97. H.L. Do, Soft-switching SEPIC converter with ripple-free input current. IEEE Trans. Power Electron. **27**(6), 2879–2887 (2011)
98. R. Haroun, A. El Aroudi, A. Cid-Pastor, G. Garcia, C. Olalla, L. Martinez-Salamero, Impedance matching in photovoltaic systems using cascaded boost converters and sliding-mode control. IEEE Trans. Power Electron. **30**(6), 3185–3199 (2014)
99. B. Tamyurek, B. Kirimer, An interleaved high-power flyback inverter for photovoltaic applications. IEEE Trans. Power Electron. **30**(6), 3228–3241 (2014)
100. A. El Khateb, N. Abd Rahim, J. Selvaraj, M.N. Uddin, B.W. Williams, Fuzzy-logic-controller based SEPIC converter for maximum power point tracking. IEEE Trans. Ind. Electron. **50**(4), 2349–2358 (2014)
101. M. Quamruzzaman, N. Mohammad, M.A. Matin, M.R. Alam, Highly efficient maximum power point tracking using DC–DC coupled inductor single-ended primary inductance converter for photovoltaic power systems. Int. J. Sustain. Energy **35**(9), 914–932 (2016)
102. J. Liao, J. Su, L. Chang, A single-phase transformer-less inverter with active decoupling, in *2014 IEEE 5th International Symposium on Power Electronics for Distributed Generation Systems (PEDG)*, pp. 1–6 (2014)
103. K. Ishaque, Z. Salam, A deterministic particle swarm optimization maximum power point tracker for photovoltaic system under partial shading condition. IEEE Trans. Ind. Electron. **60**(8), 3195–3206 (2012)
104. N.A. Gounden, S.A. Peter, H. Nallandula, S. Krithiga, Fuzzy logic controller with MPPT using line-commutated inverter for three-phase grid-connected photovoltaic systems. Renew. Energy **34**(3), 909–915 (2009)
105. F. Liu, S. Duan, F. Liu, B. Liu, Y. Kang, A variable step size INC MPPT method for PV systems. IEEE Trans. Ind. Electron. **55**(7), 2622–2628 (2008)
106. Y. Nishida, N. Aikawa, S. Sumiyoshi, H. Yamashita, H. Omori, A novel type of utility-interactive inverter for photovoltaic system, in *The 4th International Power Electronics and Motion Control Conference, 2004 (IPEMC 2004)*, vol. 3 (2004), pp. 1785–1790
107. S. Funabiki, T. Tanaka, T. Nishi, A new buck-boost-operation-based sinusoidal inverter circuit, in *2002 IEEE 33rd Annual IEEE Power Electronics Specialists Conference. Proceedings (Cat. No. 02CH37289)*, vol. 4 (2002), pp. 1624–1629
108. Deutsches Institut fur Normung (DIN), Verband der Elektrotechnik Elektronik Informationstechnik (VDE), *Automatic Disconnection Device between a Generator and the Public Low Voltage Grid*; VDE V 0126-1-1 2006-02 (Deutsches Institut fur Normung (DIN), Berlin, Germnay; Verband der Elektrotechnik Elektronik Informationstechnik (VDE), Frankfurt, Germany 2006)
109. International Electrotechnical Commission, Photovoltaic (PV) System, *Characteristics of the Utility Interface*; IEC Stand. 61727 Ed20 (International Electrotechnical Commission, Geneva, Switzerland, 2004)
110. R. Zubair, A.B. Ibrahim, M. Subhas, Multiinput DC–DC converters in renewable energy applications–an overview. Renew. Sustain. Energy Rev. **41**, 521–539 (2015)

111. M.R. Feyzi, S.A.K.M. Niapour, F. Nejabatkhah, S. Danyali, A. Feizi, Brushless DC motor drive based on multi-input DC boost converter supplemented by hybrid PV/FC/battery power system, in *Proceedings of the 24th Canadian Conference on Electrical and Computer Engineering (CCECE)* (Niagara Falls, 2011) pp. 442–446
112. H. Al-Atrash, T. Feng, I. Batarseh, Tri-modal half-bridge converter topology for three-port interface. IEEE Trans. Power Electron. **22**, 341–345 (2007)
113. M.R. Banaei, H. Ardi, R. Alizadeh, A. Farakhor, Non-isolated multi-input-single output DC/DC converter for photovoltaic power generation systems. IET Power Electron **7**, 2806–2816 (2014)
114. H. Matsuo, K. Kobayashi, Y. Sekine, M. Asano, W. Lin, Novel solar cell power supply system using the multiple-input DC–DC converter, in *Proceedings of the 20th International Telecommunications Energy Conference* (San Francisco, 1998) pp. 797–802
115. H. Matsuo, L. Wenzhong, F. Kurokawa, T. Shigemizu, N. Watanabe, Characteristics of the multiple-input DC–DC converter. IEEE Trans. Ind. Electron. **51**, 625–631 (2004)
116. Z. Zhang, O.C. Thomsen, M.A.E. Andersen, H.R. Nielsen, A novel dual-input isolated current-fed DC–DC converter for renewable energy system, in *Proceedings of the 26th Annual IEEE Applied Power Electronics Conference and Exposition (APEC)* (Fort Worth, 2011) pp. 1494–1501
117. Z. Zhang, O.C. Thomsen, M.A. Andersen, H.R. Nielsen, Dual-input isolated fullbridge boost dc–dc converter based on the distributed transformers. IET Power Electron. **5**, 1074–1083 (2012)
118. Y.M. Chen, Y.C. Liu, F.Y. Wu, T.F. Wu, Multi-input DC/DC converter based on the flux additivity, in *Proceedings of the 36th IAS Annual Meeting Industry Applications Conference*, vol. 3 (Chicago, 2001), pp. 1866–1873
119. C. Yaow-Ming, L. Yuan-Chuan, W. Feng-Yu, Multi-input DC/DC converter based on the multiwinding transformer for renewable energy applications. IEEE Trans. Ind. Electron. **38**, 1096–1104 (2002)
120. W. Hongfei, X. Yan, C. Runruo, Z. Junjun, S. Kai, G. Hongjuan, A three-port half-bridge converter with synchronous rectification for renewable energy application, in *IEEE Energy Conversion Congress and Exposition (ECCE)* (Phoenix, 2011), pp. 3343–3349
121. W. Hongfei, C. Runruo, Z. Junjun, X. Yan, H. Haibing, G. Hongjuan, A family of three-port half-bridge converters for a stand-alone renewable power system. IEEE Trans. Power Electron. **26**, 2697–2706 (2011)
122. Y.M. Chen, Y.C. Liu, S.H. Lin, Double-input PWM DC/DC converter for high/low voltage sources, in *Proceedings of the 25th International Telecommunications Energy Conference* (Yokohama, Japan, 2003) pp. 27–32
123. K.P. Yalamanchili, M. Ferdowsi, K. Corzine, New double input DC–DC converters for automotive applications, in *Proceedings of the IEEE Vehicle Power and Propulsion Conference* (Windsor, 2006) pp. 1–6
124. C. Yaow-Ming, L. Yuan-Chuan, L. Sheng-Hsien, Double-input PWM DC/DC converter for high-/low-voltage sources. IEEE Trans. Ind. Electron. **53**, 1538–1545 (2006)
125. K. Gummi, M. Ferdowsi, Double-input DC–DC power electronic converters for electric-drive vehicles-topology exploration and synthesis using a single pole triple-throw switch. IEEE Trans. Ind. Electron. **57**, 617–623 (2010)
126. L. Solero, F. Caricchi, F. Crescimbini, O. Honorati, F. Mezzetti, Performance of a 10 kW power electronic interface for combined wind/PV isolated generating systems, in *Proceedings of the 27th Annual IEEE Power Electronics Specialists Conference*, vol. 2 (Baveno, 1996) pp. 1027–1032
127. F.D. Rodriguez, W.G. Imes, Analysis and modeling of a two-input DC/DC converter with two controlled variables and four switched networks, in *Proceedings of the 31st Intersociety Energy Conversion Engineering Conference*, vol. 1 (Washington, 1996). pp. 322–327
128. A. Di Napoli, F. Crescimbini, S. Rodo, L. Solero, Multiple input DC–DC power converter for fuel-cell powered hybrid vehicles, *in Procceedings of the IEEE 33rd Annual Power Electronics Specialists Conference* (Cairns, 2002) pp. 1685–1690

129. L. Solero, A. Lidozzi, J.A. Pomilio, Design of multiple-input power converter for hybrid vehicles. in *Proceedings of the 19th Annual IEEE Applied Power Electronics Conference And Exposition*, vol. 2 (2004) pp. 1145–1151
130. L. Solero, A. Lidozzi, J.A. Pomilio, Design of multiple-input power converter for hybrid vehicles. IEEE Trans. Power Electron. **20**, 1007–1016 (2005)
131. J. Zhang, H. Wu, J. Huang, Y. Xing, X. Ma, A novel multi-port bidirectional converter for interfacing distributed DC micro-grid, in *Proceedings of the IEEE 23rd International Symposium on Industrial Electronics (ISIE)* (Istanbul, 2014) pp. 2344–2348
132. H. Tao, A. Kotsopoulos, J.L. Duarte, M.A.M. Hendrix, Multi-input bidirectional DC-DC converter combining DC-link and magnetic-coupling for fuel cell systems vol. 3, in Proceedings of the 40th IAS Annual Meeting Industry Applications Conference (2005), pp. 2021–2028

Mission Profile Oriented Reliability Evaluation of Grid-Connected PV Inverter Considering Panel Degradation and Uncertainties at Indian Location

Sainadh Singh Kshatri, Javed Dhillon, and Sachin Mishra

1 Introduction

In the recent decade, grid penetration of solar PV energy systems gained more popularity due to its potential and advancements in the PV technology [1]. Researchers reported that among the renewable energy sources, PV energy has the potential to share the significant amount of total electricity demand [2–4]. The advancements of inverters (plug and play installations) and array-level MPPT units increase the PV energy yield in uncertain environmental conditions. At the end of 2019, total global PV installed capacity is reported as 627 GW, China ranks first with 204.7 GW, India at fifth position with 42.8 GW. At least 18 countries have the PV installed capacity more than 1 GW, 9 countries have more than 10 GW. Germany reported the highest solar per capita, i.e., 595 followed by Australia and Japan. Top ten countries for annual installed capacities and cumulative capacities are shown in Figs. 1 and 2, respectively [5].

The 2019 India stands at third position for annual installed capacity, this shows the potential of PV market in India. Nevertheless, reliability performance of the PV system is a major concern. In [6], it is stated that 59% of the cost associated with PV system is inverter hence reliability of PV inverter has a significant impact. Environmental factors like mission profile (solar irradiance, ambient temperature), panel degradation affect the reliability performance of PV inverter [7–9]. A survey in [10] states that 31% of critical component in the PV inverter is power electronic switches as shown in Fig. 3.

S. S. Kshatri (✉) · J. Dhillon · S. Mishra
School of Electronics & Electrical Engineering, Lovely Professional University, Phagwara, Punjab, India
e-mail: sainadhsingh@gmail.com

S. S. Kshatri
Department of Electrical & Electronics Engineering, B V Raju Institute of Technology, Narsapur, India

V. K. Singh et al. (eds.), *Renewable Energy and Future Power Systems*, Energy Systems in Electrical Engineering, https://doi.org/10.1007/978-981-33-6753-1_3

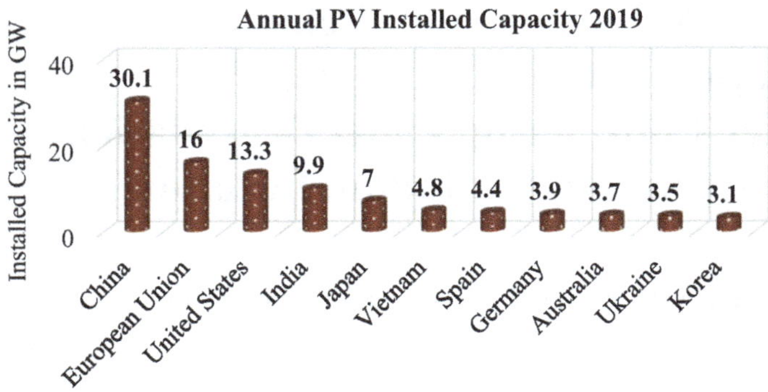

Fig. 1 Top ten countries annual PV installed capacity

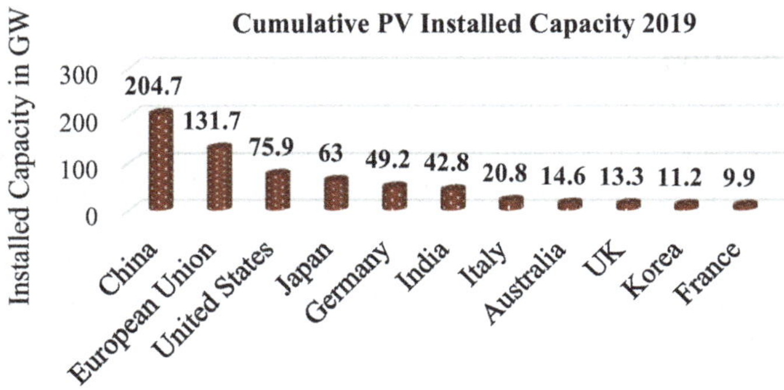

Fig. 2 Top ten countries cumulative PV installed capacity

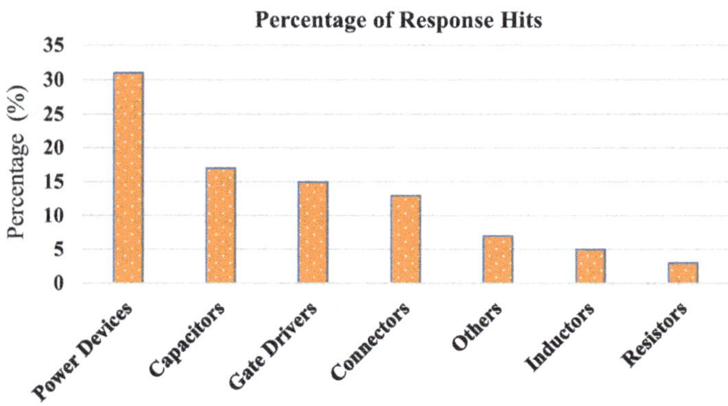

Fig. 3 Percentage of critical component of PV inverter

Table 1 Critical stressors of IGBT

Types of load		IGBT		
Climate condition	Stressor	Die	Solder joints	Wire-bond
Temperature	Temperature swing	High	High	High
	Average temperature	Medium	Medium	Medium
	dT/dt	Low	Low	Low
Shock	Mechanical shock	Low	Neglectable	Neglectable
Humidity/Moisture	Relative humidity	Low	Low	Low
Pollution	Pollution	Neglectable	Neglectable	Neglectable

In [11], it is stated that the critical stressor of Power Electronic switch (IGBT) is thermal stress and tabulated in Table 1.

According to [11], there are mainly two types of IGBT failures: bond-wire lift off and die-attach solder fatigue [6], [12–16]. Hence, reliability of IGBT needs to be assessed. In this chapter, reliability evaluation of PV inverter is performed. A case study of 3-kW grid-connected full-bridge PV inverter with four 600 V/30A IGBT is considered. Real-time mission profile data for one year with one-minute sample is logged at B V Raju institute of Technology, Narsapur, Medak, India. Two-parameter Weibull distribution based reliability evaluation is carried out, both component-level and system-level PV inverter reliabilities are analyzed. Results reveal that mission profile, panel degradation rate, and uncertainties significantly impact the reliability performance of PV inverter.

2 Reliability Evaluation of Grid-Connected PV Inverter

Reliability evaluation flowchart of PV inverter consists of several stages. In stage I, mission profile for one year at India location needs to be logged as shown in Fig. 4. Pyranometer is used to record solar irradiance, RTD device is used to record ambient temperature, CR3000 data logger is used to log the data with one-minute resolution.

In stage II, junction temperature (T_j) of IGBT is calculated with the help of Foster electro-thermal model as shown in Fig. 5.

In stage III, rainflow counting algorithm is used to analyze the variations of T_j. From the rain flow counting algorithm Number of Cycles n_i, Mean Junction Temperature T_{jm}, Cycle Amplitude ΔT are calculated [17]. In stage IV, lifetime is calculated using Miner's rule as shown below [18].

$$\text{Life Time}\,(LT) = \frac{1}{\text{Life Consumption}\,(LC)} \tag{1}$$

Life Consumption of IGBT is given as

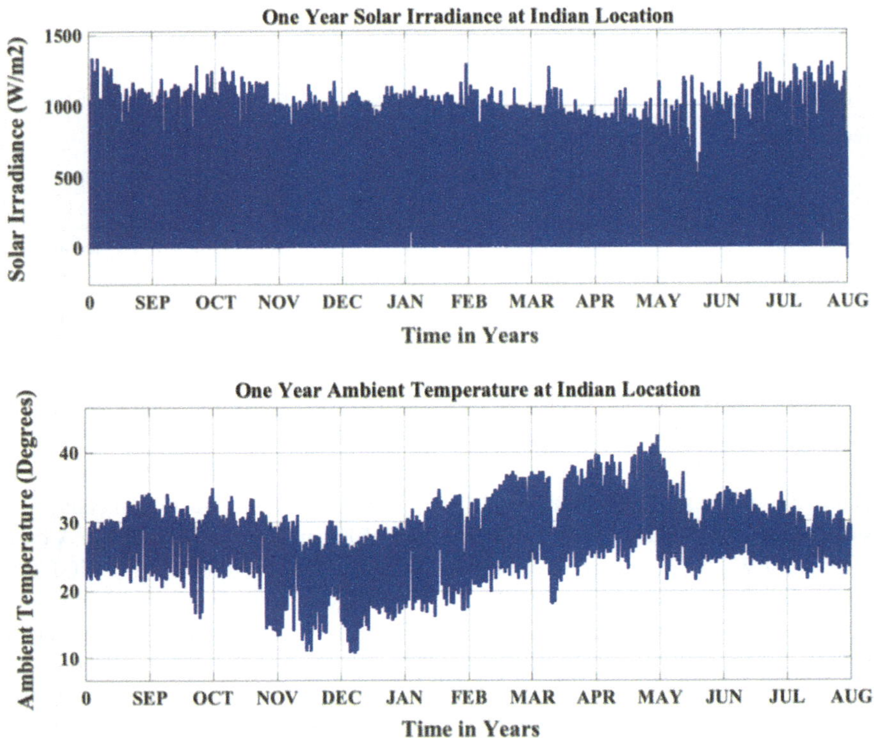

Fig. 4 One-year mission profile

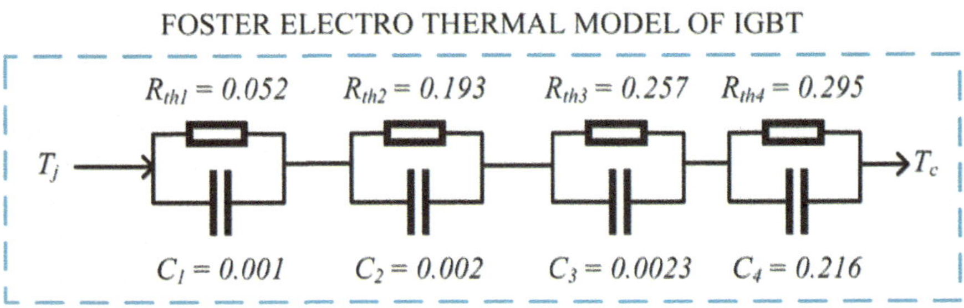

Fig. 5 Foster electro-thermal model of IGBT

$$\text{Life Consumption}\,(LC) = \sum \frac{\text{No. of Cycles}\,(n_i)}{\text{No. of Cycle to failure}\,\left(N_{fi}\right)} \tag{2}$$

$$N_f = K\left(\Delta T_j\right)^{\beta_1}.e^{\frac{\beta_2}{\left(T_j+273K\right)}}.t_{on}^{\beta_3}.I^{\beta_4}.V^{\beta_5}.D^{\beta_6} \tag{3}$$

N_f is taken from Bayerer's lifetime model. The parameters of Eq. 3 are tabulated in Table 2

Table 2 Bayerer's lifetime model parameters

Parameter	Value
Technology Factor (A)	9.34E + 14
Coefficient (β_1)	−4.416
Coefficient (β_2)	1285
Coefficient (β_3)	−0.463
Coefficient (β_4)	−0.716
Coefficient (β_5)	−0.761
Coefficient (β_6)	−0.5
Current per bond foot (I)	3–23 A
Voltage class (V)	6–33 V
Bond wire diameter	75-500 μm

In stage 10000 sample with five percent variations (Uncertainty) are generated using Monte Carlo simulation and fitted in two-parameter Weibull distribution. The reliability function is given as

For component level, reliability function is given as

$$R(t) = e^{-\left(\frac{t}{\alpha}\right)^{\gamma}} \tag{4}$$

For system level,

$$R_{total}(t) = \prod_{i=1}^{n} R_i(t) \tag{5}$$

For the Weibull reliability function B_x lifetime is calculated. The reliability flow is shown in Fig. 6 [19–23].

3 Results and Discussion

In this chapter 3-kW grid-connected full-bridge PV inverter with four 600 V/30A IGBT is considered as shown in Fig. 7. The proposed system is evaluated under two cases.

- Reliability Evaluation of PV Inverter without Degradation Rate.
- Reliability Evaluation of PV Inverter with Degradation Rate.

The parameters of the proposed system are tabulated in Table 3.

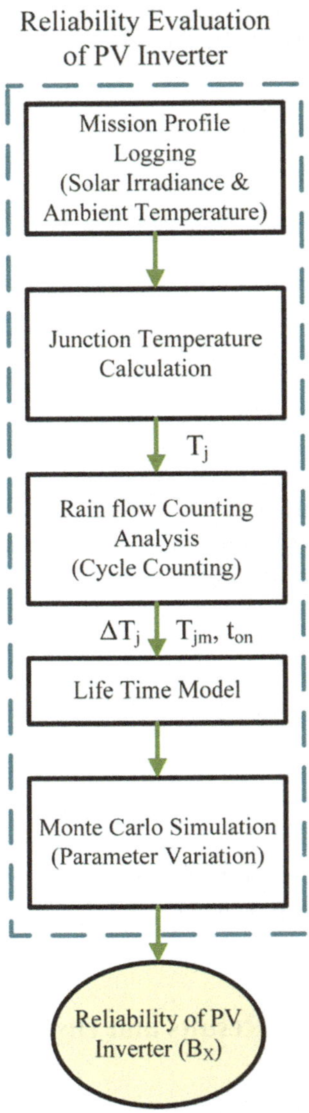

Fig. 6 Reliability evaluation flowchart

3.1 Reliability Evaluation of PV Inverter Without Degradation Rate

In this case, reliability evaluation of PV inverter is considered without degradation rate.

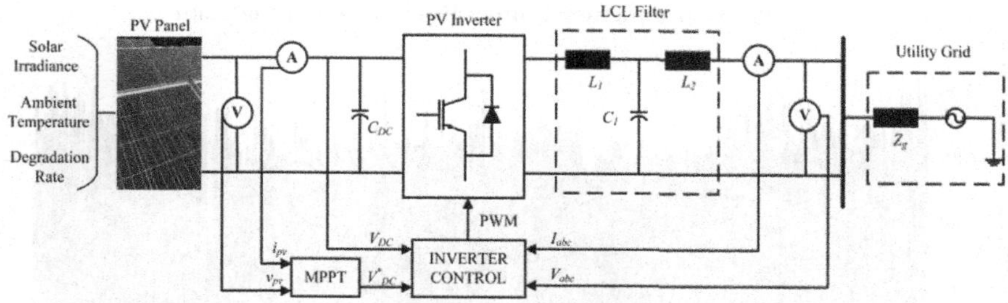

Fig. 7 3-kW grid-connected PV system

Table 3 Parameters of the Proposed System

Name	Specification
PV panel model	BP365
PV inverter rated power	3 kW
Grid voltage	230 V
Grid frequency	50 Hz
Dc link capacitance	1.5e–3 F
IGBT make	Infineon
IGBT model	IGW30N60H3
IGBT Voltage rating	600 V
IGBT Current rating	30 A
IGBT min T_j	25 °C
IGBT max T_j	175 °C

3.1.1 Estimation of Junction Temperature Corresponds to Mission Profile

The junction temperature of 600 V/30A IGBT is estimated using Foster electro-thermal model corresponds to yearly mission profile at Indian location is as shown in Fig. 8.

The junction temperature T_j recorded maximum is 103.8°C, minimum is 11.4°C. It is continuously varying in correlation with mission profile. In contrast, a cycle counting algorithm is needed to analyze the variations.

3.1.2 Rainflow Counting Algorithm

The rainflow counting algorithm translates the irregular junction temperature of PV inverter without degradation rate into regular profiles. It calculates the number of cycles i.e. temperature variations at the junction layers of IGBT. It regularizes the junction temperature and extracts Mean Junction Temperature T_{jm}, Cycle Amplitude

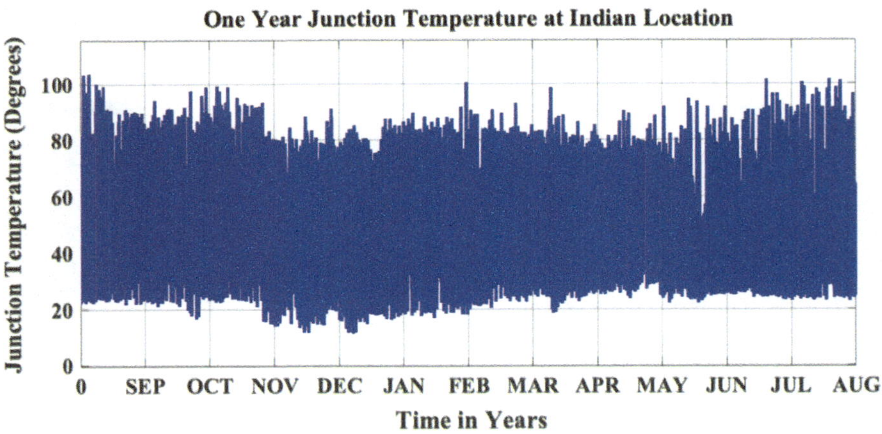

Fig. 8 One-year junction temperature at Indian location

Table 4 Static rainflow parameters

Static rainflow parameters	Value
Number of Cycles n_i,	33090
Mean Junction Temperature T_{jm}	5.94 °C
Cycle Amplitude ΔT	59.77 °C

ΔT at each cycle. The static rainflow parameters are shown in Table 4. Rainflow histogram, Cycle Amplitude ΔT, Mean Junction Temperature T_{jm} are as shown in Fig. 9a–c.

3.1.3 Lifetime Estimation

The lifetime is calculated using Eq. 1 considering static values of rainflow parameters i.e. all devices fail at same rate. Lifetime for static values is tabulated in Table 5.

The above lifetime is estimated for single IGBT.

3.1.4 Reliability Evaluation Using Monte Carlo Simulation

In the previous stage lifetime is calculated considering all the devices fails at same rate but practically never fails at same rate. Hence 10000 samples with five percent variations (Uncertainty) are generated using Monte Carlo simulation and fitted in two-parameter Weibull distribution. The lifetime distribution of PV inverter is shown in Fig. 10.

The reliability of PV inverter without degradation rate is shown in Fig. 11.

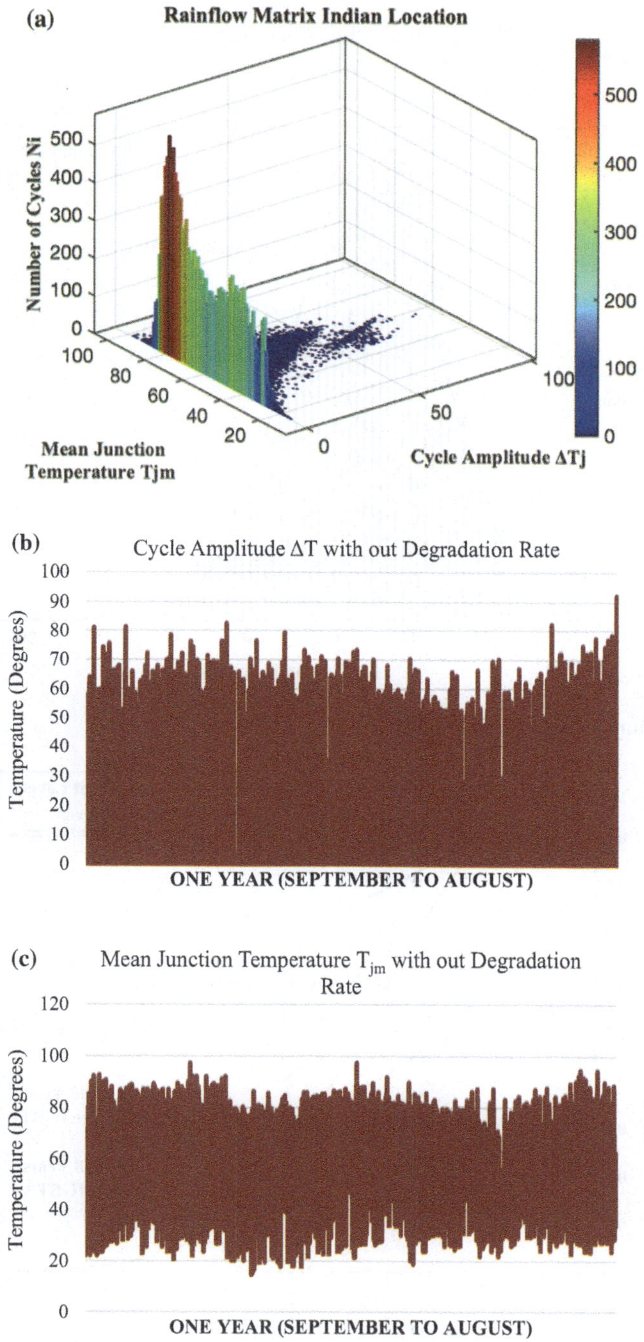

Fig. 9 **a** Rainflow Histogram. **b** Cycle Amplitude ΔT without Degradation Rate. **c** Mean Junction Temperature T_{jm} without Degradation Rate

Table 5 Lifetime Estimation for static parameters

Country	Life Consumption (LC)	Lifetime in Years (LF)
India	0.00908	110.06

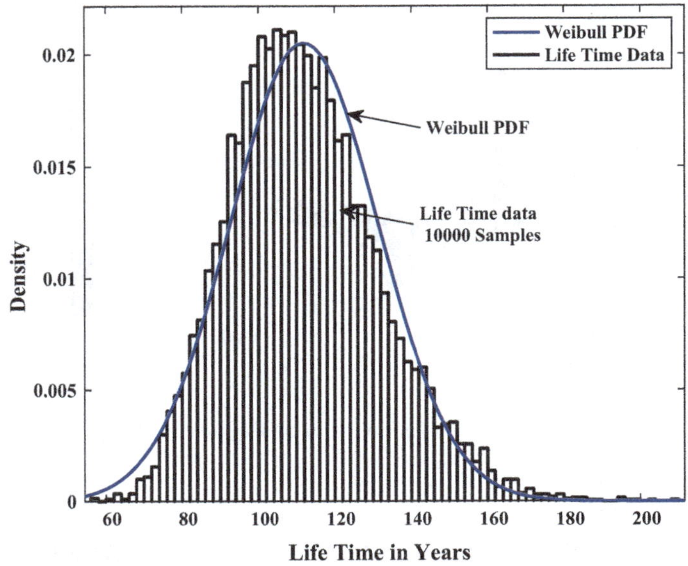

Fig. 10 Lifetime distribution of PV Inverter

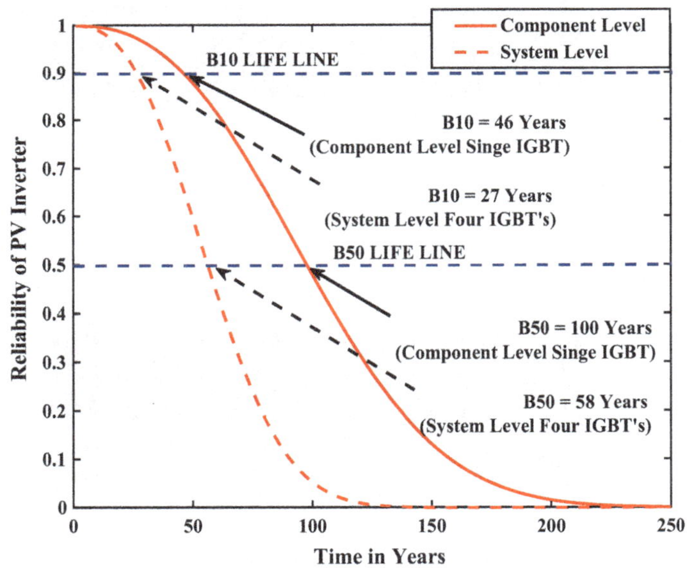

Fig. 11 Reliability of PV Inverter without degradation rate

The B_{10} lifetime of PV inverter at component level is 46 years, at system level is 27 Years. Similarly, B_{50} lifetime of PV inverter at component level is 100 years, at system level is 58 Years

3.2 Reliability Evaluation of PV Inverter with Degradation Rate

In this case reliability evaluation of PV inverter is considered with degradation rate. Degradation rate of 1.93% is considered at Indian location from [].

3.2.1 Estimation of Junction Temperature Corresponds to Mission Profile

The junction temperature of 600 V/30A IGBT is estimated using foster electro thermal model corresponds to yearly mission profile at Indian location is as shown in Fig. 12.

The junction temperature T_j recorded maximum is 102.5 °C, minimum is 10.3 °C. It is continuously varying in correlation with mission profile. In contrast a cycle counting algorithm is needed to analyze the variations.

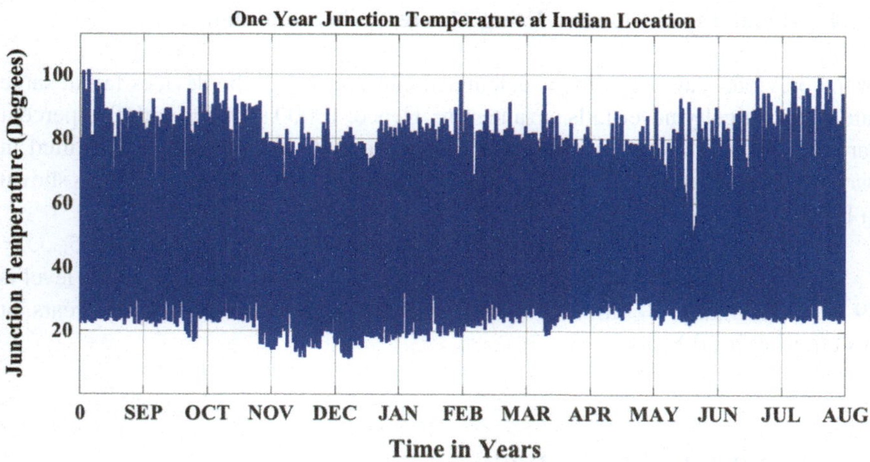

Fig. 12 One-year Junction Temperature at Indian Location

Table 6 Static rainflow parameters

Static rainflow parameters	Value
Number of Cycles n_i,	33092
Mean Junction Temperature T_{jm}	5.92 °C
Cycle Amplitude ΔT	59.67 °C

3.2.2 Rainflow Counting Algorithm

The rainflow counting algorithm translates the irregular junction temperature of PV inverter with degradation rate into regular profiles. It calculates the number of cycles, i.e., temperature variations at the junction layers of IGBT. It regularizes the junction temperature and extracts Mean Junction Temperature T_{jm}, Cycle Amplitude ΔT at each cycle. The static rainflow parameters are shown in Table 6. Rainflow histogram, Cycle Amplitude ΔT, Mean Junction Temperature T_{jm} are as shown in Fig. 13a–c

3.2.3 Lifetime Estimation

The lifetime is calculated using Eq. 1 considering static values of rainflow parameters, i.e., all devices fail at same rate. Lifetime for static values is tabulated in Table 7.

The above lifetime is estimated for single IGBT.

3.2.4 Reliability Evaluation Using Monte Carlo Simulation

In the previous stage, lifetime is calculated considering all the devices fail at same rate but practically never fails at same rate. Hence, 10000 samples with five percent variations (Uncertainty) are generated using Monte Carlo simulation and fitted in two-parameter Weibull distribution. The lifetime distribution of PV inverter is shown in Fig. 14.

The reliability of PV inverter without degradation rate is shown in Fig. 15.

The B_{10} lifetime of PV inverter at component level is 48 years, at system level is 29 Years. Similarly, B_{50} lifetime of PV inverter at component level is 102 years, at system level is 60 Years.

3.3 *B_x Lifetime Comparison*

In this chapter, B_x lifetime is calculated considering with and without degradation rate. The panel degradation leads to increase in lifetime, i.e., thermal loading on PV inverter decreases. B_x lifetime comparison is shown in Fig. 16.

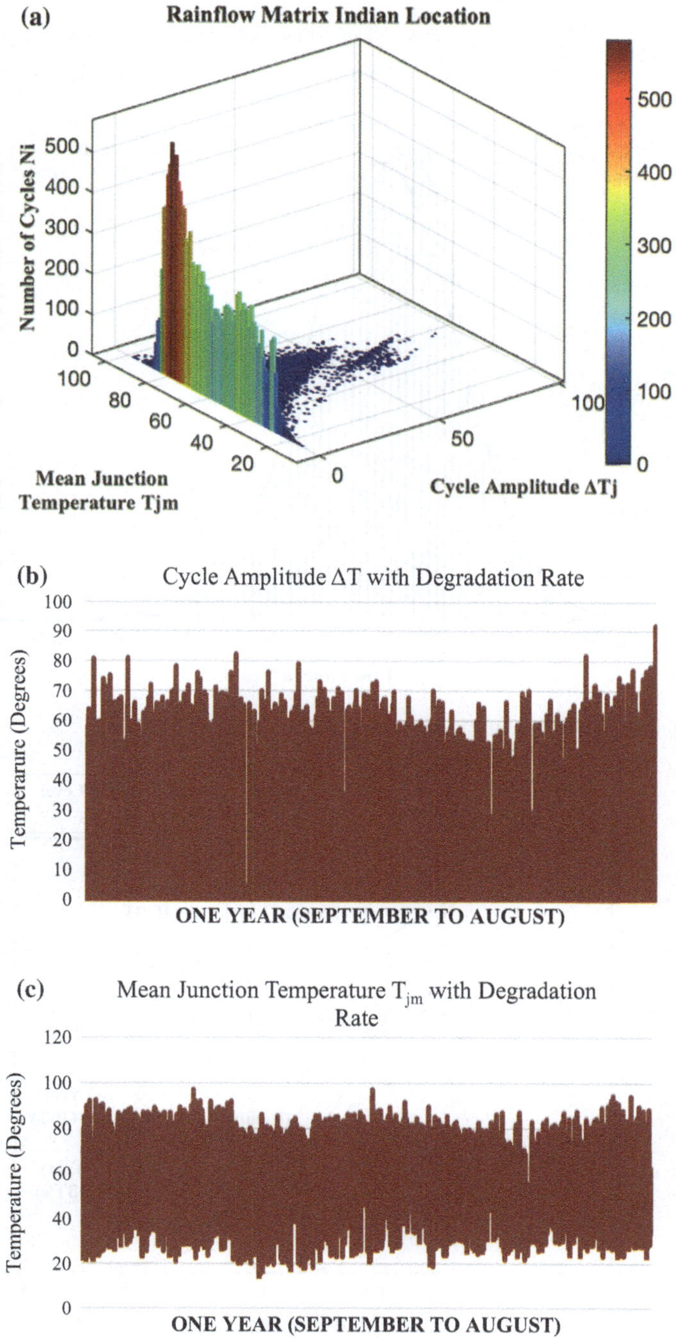

Fig. 13 **a** Rainflow Histogram. **b** Cycle Amplitude ΔT with Degradation Rate. **c** Mean Junction Temperature T_{jm} with Degradation Rate

Table 7 Lifetime estimation for static parameters

Country	Life Consumption (LC)	Lifetime in Years (LF)
India	0.00891	112.125245

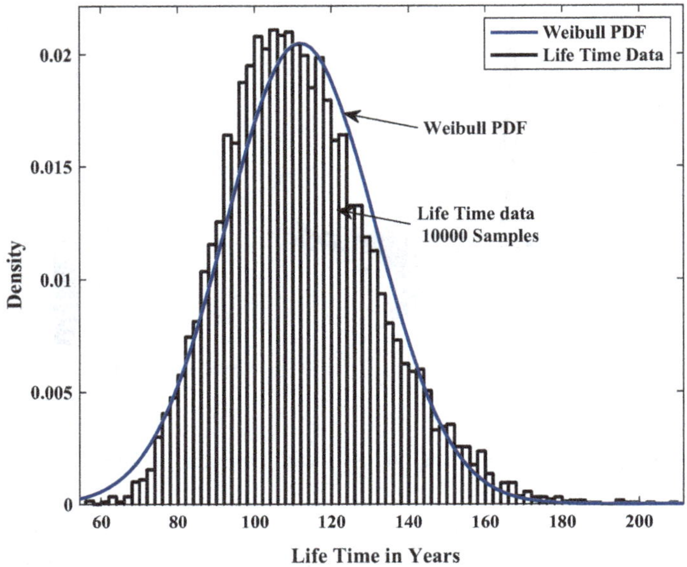

Fig. 14 Lifetime distribution of PV Inverter

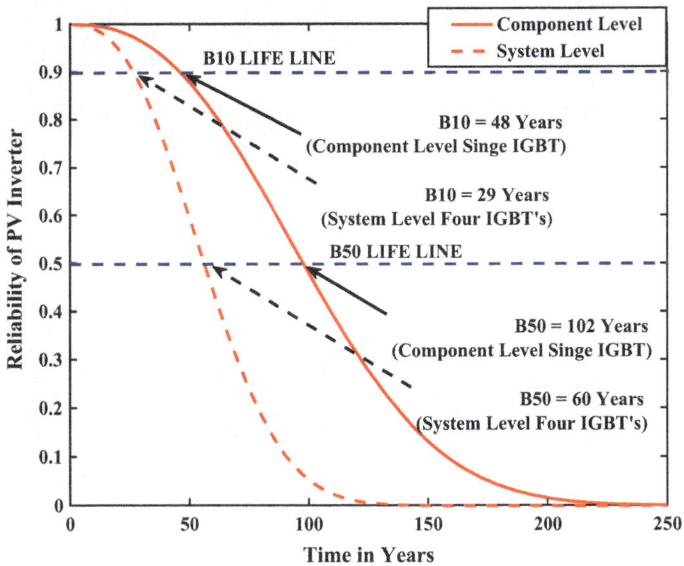

Fig. 15 Reliability of PV Inverter without degradation rate

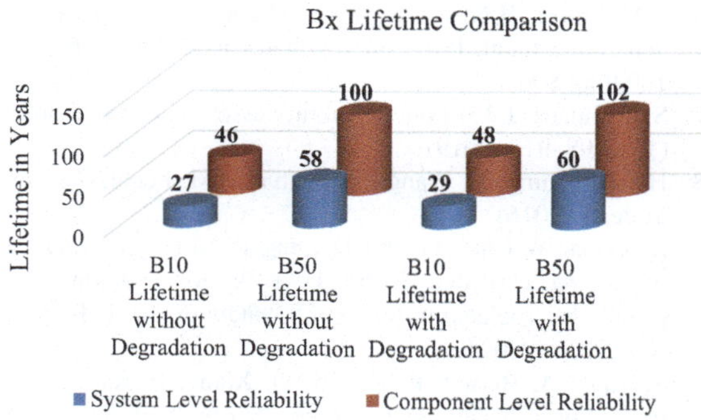

Fig. 16 B_x lifetime comparison

4 Conclusion

This chapter presents the impact of mission profile, panel degradation, and uncertainties on PV inverter reliability. A case study of 3-kW grid-connected full-bridge PV inverter with four 600 V/30A IGBT is considered for reliability evaluation. Real-time mission profile data for one year with one-minute sample is logged at B V Raju institute of Technology, Narsapur, Medak, India. Foster electro-thermal model is used for the junction temperature estimation of IGBT. Rainflow counting algorithm is used to analyze the junction temperature variations. 10000 samples with five percent variations (Uncertainty) are generated using Monte Carlo simulation. Two-parameter Weibull distribution based reliability evaluation is carried out, both component level and system level PV inverter reliability is analyzed. Results reveal that mission profile, panel degradation rate and uncertainties significantly impact the reliability performance of PV inverter. B_{10}, B_{50} lifetime with and without degradation rate are calculated and compared.

References

1. R. Panigrahi, S.K. Mishra, S.C. Srivastava, A.K. Srivastava, N.N. Schulz, Grid integration of small-scale photovoltaic systems in secondary distribution network—a Review. IEEE Trans. Ind. Appl. **56**(3), 3178–3195 (2020). https://doi.org/10.1109/TIA.2020.2979789
2. A. Sangwongwanich, Y. Yang, D. Sera, F. Blaabjerg, Impacts of PV array sizing on PV inverter lifetime and reliability, *2017 IEEE Energy Conversion Congress and Exposition (ECCE)*, vol. 2017 (2017) pp. 3830–3837. https://doi.org/10.1109/ecce.2017.8096675
3. Y. Yang, A. Sangwongwanich, F. Blaabjerg, Design for reliability of power electronics for grid-connected photovoltaic systems. CPSS Trans. Power Electron. Appl. 1(1) (2016)
4. A. Sangwongwanich, Y. Yang, D. Sera, F. Blaabjerg, Lifetime evaluation of grid-connected PV inverters considering panel degradation rates and installation sites. IEEE Trans. Power Electron. **33**(2), 1125–1236 (2018). https://doi.org/10.1109/TPEL.2017.2678169
5. IEA-Pvp, Snapshot of Global PV Markets (2020) Www.Iea-Pvps.Org, http://www.iea-pvps.org/fileadmin/dam/public/report/technical/PVPS_report_-_A_Snapshot_of_Global_PV_-_1992-2014.pdf, pp. 1–16

6. L.M. Moore, H.N. Post, Five years of operating experience at a large, utility-scale photovoltaic generating plant. Prog. Photovoltaics Res. Appl. **16**(3), 249–259 (2008). https://doi.org/10.1002/pip.800
7. S.S. Kshatri, J. Dhillon, Reliability assessment methodologies for grid connected PV inverter (17), 908–917 (2019)
8. H.S.H. Chung, H. Wang, F. Blaabjerg, M. Pecht, *Reliability of Power Electronic Converter Systems* (2016)
9. R. Kumar, W. Chen, G. Zhu, D. Zeng, R. Nirudi, Microelectronics reliability lifetime estimation of modular cascaded H-bridge MLPVI for grid-connected PV systems considering mission profile. Microelectron. Reliab. **88–90**(June), 1051–1056 (2018). https://doi.org/10.1016/j.microrel.2018.06.024
10. S. Yang, A. Bryant, P. Mawby, D. Xiang, L. Ran, P. Tavner, An industry-based survey of reliability in power electronic converters. IEEE Trans. Ind. Appl. **47**(3), 1441–1451 (2011). https://doi.org/10.1109/TIA.2011.2124436
11. W. Huai et al., Transitioning to physics-of-failure as a reliability driver in power electronics. IEEE J. Emerg. Sel. Top. Power Electron. **2**(1), 97–114 (2014). https://doi.org/10.1109/JESTPE.2013.2290282
12. E. Liivik, A. Sangwongwanich, F. Blaabjerg, Reliability analysis of micro-inverters considering PV module variations and degradation rates, in *2018 20th European Conference on Power Electronics and Applications (EPE'18 ECCE Europe)* (2018), pp. P1–P8
13. A. Sangwongwanich et al., Reliability assessment of PV inverters with battery systems considering PV self-consumption and battery sizing, in *2018 IEEE Energy Conversion Congress and Exposition (ECCE)* (2018), pp. 7284–7291
14. A. Sangwongwanich et al., Enhancing PV inverter reliability with battery system control strategy. CPSS Trans. Power Electron. Appl. **3**(2), 93–101 (2018)
15. S.G. Kadwane, J.M. Kumbhare, D.K. Mohanta, A Matlab/Simulink modeling for reliability analysis of inverter applied to MPPT based PV system, in *2015 Annual IEEE India Conference (INDICON)* (2015), pp. 1–6
16. S.B. Narale, P.N. Kumar, M.W. Ahmad, D.S. More, S. Anand, Mission profile based evaluation of capacitor reliability in two stage grid feeding photovoltaic inverter, in *2017 National Power Electronics Conference (NPEC)* (2017), pp. 7–11
17. L.R. GopiReddy, L.M. Tolbert, B. Ozpineci, J.O.P. Pinto, Rainflow algorithm-based lifetime estimation of power semiconductors in utility applications. IEEE Trans. Ind. Appl. (2015). https://doi.org/10.1109/TIA.2015.2407055
18. R.K. Gatla, W. Chen, G. Zhu, J.V. Wang, S.S. Kshatri, Lifetime comparison of IGBT modules in grid-connected multilevel PV inverters considering mission profile, in *ICPE 2019 ECCE Asia–10th International Conference on Power Electronics ECCE Asia* (2019)
19. R. Kaplar et al., PV inverter performance and reliability: what is the role of the IGBT? in *2011 37th IEEE Photovoltaic Specialists Conference* (2011), pp. 1842–1847
20. Y. Shen, H. Wang, Y. Yang, P.D. Reigosa, F. Blaabjerg, Mission profile based sizing of IGBT chip area for PV inverter applications, in *2016 IEEE 7th International Symposium on Power Electronics for Distributed Generation Systems (PEDG)* (2016), pp. 1–8
21. Y. Shen, E. Liivik, F. Blaabjerg, D. Vinnikov, H. Wang, A. Chub, Reliability evaluation of an impedance-source PV microconverter, in 2018 *IEEE Applied Power Electronics Conference and Exposition (APEC)* (2018), pp. 1104–1108
22. A. Sangwongwanich, E. Liivik, F. Blaabjerg, Photovoltaic module characteristic influence on reliability of micro-inverters, in *2018 IEEE 12th International Conference on Compatibility, Power Electronics and Power Engineering (CPE-POWERENG 2018)* (2018), pp. 1–6
23. P. Zhang, Y. Wang, W. Xiao, W. Li, Reliability evaluation of grid-connected photovoltaic power systems. IEEE Trans. Sustain. Energy **3**(3), 379–389 (2012)

Performance Analysis of PI and PR Controller for a Single-Phase PV Grid System with Effective Active and Reactive Power Compensation

Srikanth Sattenapalli and V. Joshi Manohar

1 Introduction

Nowadays, because of advancements in industrialization, transportation, and utilization in various applications, the demand for electricity may increase to 85% by 2040. The conventional energy sources usually include gas, coal, nuclear, and hydel power plants are chosen [1, 2]. The main problem with utilization of these conventional plants is they cause pollution and greenhouse gas emissions and also damages the environmental conditions. To overcome this effect, a renewable energy system is chosen. Increasing the supply of nonconventional sources reduces the carbon-intensive energy sources and significantly reduce the global warming emissions.

In the present scenario, the utilization of renewable energy has been increased rapidly to meet the electrical demand [3, 4]. In this paper, a PV system is considered as input source and connected to the grid, during this interaction between grid and distributed source a power quality problem arises. Iin small-scale power generating plants, the Distributed Energy Sources play a key role, for example, nonconventional power sources like photovoltaic cell, wind systems, hydro sources designing the microgrid idea close to the load center have the advantage of improving effectiveness by decreasing the transmission line misfortunes or voltage drops.

The photovoltaic system is one of the most convenient renewable energy systems when compared to all other renewable energy sources. Photovoltaic systems are not naturally stable in time, location, season, and weather and also the cost for installation of PV system is very high [5–7]. The output of generated PV system depends on

S. Sattenapalli (✉)
Department of Electrical and Electronics Engineering, Koneru Lakshmaiah Education Foundation, Vaddeswaram, A.P, India
e-mail: zallerforu@gmail.com

V. Joshi Manohar
Department of EEE, Presidency University, Bangalore, India

V. K. Singh et al. (eds.), *Renewable Energy and Future Power Systems*, Energy Systems in Electrical Engineering, https://doi.org/10.1007/978-981-33-6753-1_4

weather conditions, where the weather condition is not constant. Therefore, to get better output and increasing the performance from the PV panel, an MPPT technique is implemented. Generally, the maximum power point position is unknown, and this can be achieved by using some searching algorithms to track power and getting efficiency. In this paper, an Incremental Conductance based MPPT technique is used. And also, the system needs to maintain synchronization with the grid. For this, to match the frequency levels and PV grid system, it is connected to voltage source inverter. The control diagram for the inverter is designed with general PWM technique and the reference signals are chosen from the grid parameters. The most significant feature of the residential solar energy systems is that electricity is produced every day. The microgrid contains a local storehouse of energy, which can be stored in a battery bench when photovoltaic panels fail to supply power [8]. The delivery functionality or feature may be linked in standalone mode directly. Microgrid allows for electricity distribution to be optimally controlled, local renewable energy output is regulated and funded, if necessary, by the local power generation [9]. Without a steady-state error caused by the integral term dynamic, the limit of the PI controller cannot follow the Sinusoidal Reference. The inability to follow a sinusoidal reference means that the door voltage is used for the anticipation of a high dynamic response by helping the controller to achieve a steady state more quickly. Otherwise, the PR regulator is ideally adapted without the latter drawback to sinusoidal comparisons. Just a frequency [10] provides a benefit. In this sense, the present paper discusses both the architecture of a photovoltaic system-connected single-phase grid and examines PI/ PR problems at the level of current regulators. This thesis ends with a comparative analysis of the two styles of controllers.

2 Grid-Interconnected Network and MPPT Technique

In general, the coordination of grid sustainable power sources, for example, photovoltaic framework is the most significant application. These points of interest incorporate the ideal motivators in numerous nations that effect clearly on the business acknowledgment of grid-associated PV frameworks [5, 6, 10]. This condition forces the need of having great quality planning apparatuses and an approach to precisely foresee the dynamic execution of three-stage grid-associated PV frameworks under various working conditions so as to settle on a cool-headed choice on whether to consolidate this innovation into the electric utility grid.

2.1 PV Solar System

In present scenario, the solar PV system plays a key role in distribution energy systems as its flexibility and reliable nature. The PV Array converts sun irradiance to electrical energy by photon effect [4, 8]. In PV system, the solar cells are arranged in series and parallel combinations to meet the load requirement such as voltage and current. A structure of general grid system is shown in Fig. 1. It represents the flow of energy in the system where Grid connection can be interacted with Renewable Energy sources through inverter and switchgear. The power conversion process is shown in Fig. 2, where MPPT controller tracks the maximum power from solar panel and is connected to Boost converter for effective power output.

The main components in solar system are described as follows (refer to Fig. 2). PV panel which converts sun photon light into electrical current and further it is converted to DC voltage with the help of electrical equivalent circuit. To reach the maximum power tracking from the solar system, an MPPT-based DC-DC converter is implemented. The purpose of MPPT technique is to track the power from the solar system. A maximum power point tracker is a basic DC to DC converter that

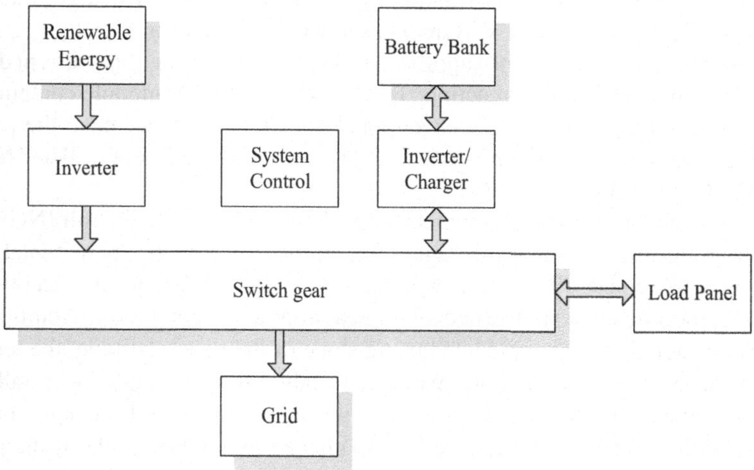

Fig. 1 Structure of general grid system

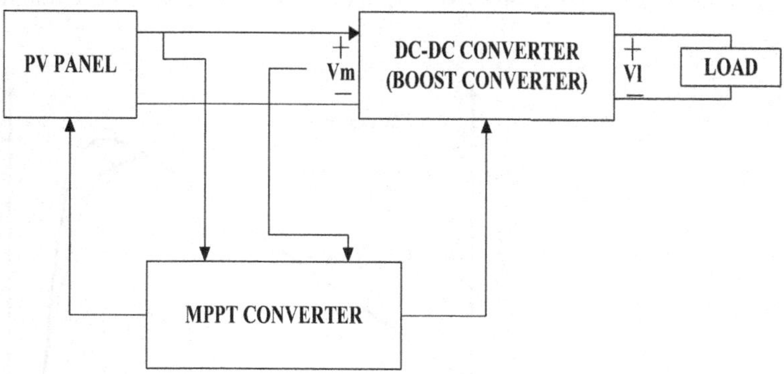

Fig. 2 PV system with power converter

synchronizes between the PV system and Point of Common Coupling. The purpose of this converter is to control the solar voltage and track the maximum power from the panels.

2.2 Incremental Conductance Method

Incremental conductance method is one of the better searching algorithms [11] in MPPT techniques. In this method, the maximum power is obtained by the slope derivative of the PV current with respect to PV voltage. The relation between the PV system voltage and current is in exponential form and the curve is in nonlinear form, so that MPPT is present at knee of the curve as shown in Fig. 3. This V-I and P-V characteristic curve gives the maximum power that can tracked with algorithm. The INC searching algorithm for PV system is shown in flowchart in Fig. 4.

The Incremental Conductance approach utilizes source voltage and current details to determine the target point of action. The P-V curve of a PV module contains two standalone control loops for the conventional MPPT Systems to make the panel's operating point the maximum power point. The second is the PI controller. The first is the MPPT algorithm.

Instantaneous, incremental conductance is the foundation for the full INC Algorithm to produce a zero error message at the point of MPP. The Incremental Lead algorithm is extremely effective since it has static state consistency and can thus be effectively modified in response to evolving environmental conditions. Additionally, if the Solar radiation is also modeled using slope of the panel, latitude, the temperature of the place where the solar panel is installed the panel can be installed in such a way it can produce maximum power throughout the day. If change in incremental conductance is zero, then there is no change in the duty cycle. If the power to voltage derivative is more than zero, then duty cycle has to be increased and for less than zero the duty cycle has to be decreased. Then the duty cycle is adjusted by the scaling factor and the decrement and increment in the duty cycle is decided by power to voltage slope. P&O and INC [8] approaches are used widely due to ease

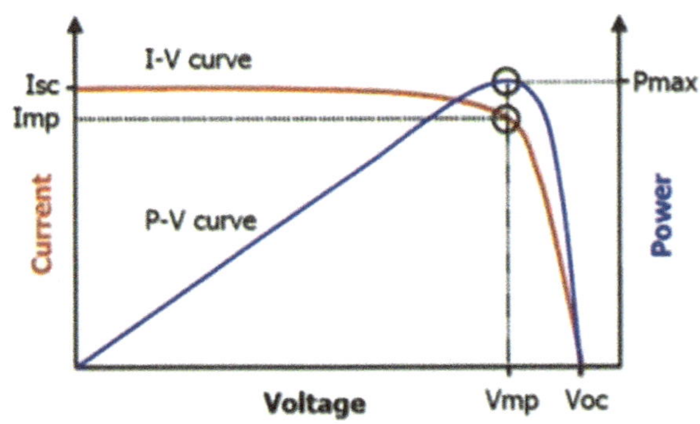

Fig. 3 P-V curve for solar module

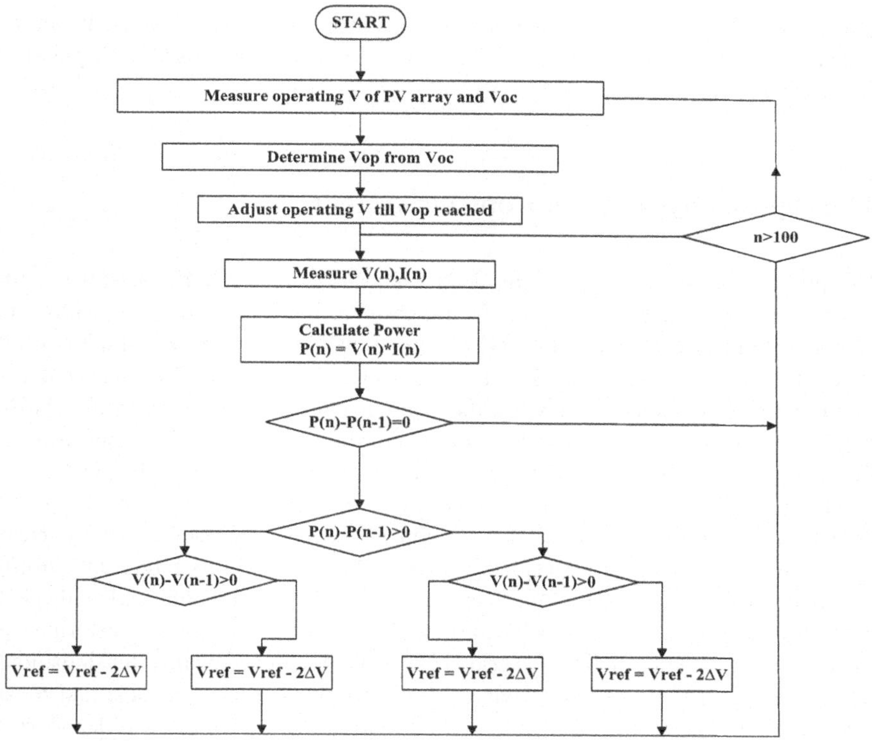

Fig. 4 Incremental conductance method algorithm

of digital implementation, independent of program parameters. P&O Algorithms [4, 12] are disadvantaged in tracking changes in the solar irradiance and more steady-state oscillations around MPP that increase energy loss in the PV system. To tackle these concerns, the INC approach is used to enhance the dynamic efficiency and precision in rapidly evolving irradiation conditions.

The maximum power point (MPP) is tracked under rapidly changing conditions by the increased-induction algorithm [7]. Incremental and instant conducting of the PV module is a priority for Lt [08]. The algorithm calculates the PV generator output power and its derivative as voltage function from measurements of voltage V and current I. The point of usage is at MPP, the point of activity on the left side of MPP and I is the PV plate, V is the voltage for the photovoltaics, and D is the cumulative operating time. A extremely accurate, stability-based tracking, and adaptability for rapidly evolving atmospheric environment is the algorithm that has a stronger output than other technologies [13, 14]. The standard incremental conductance algorithm uses usually a fixed iteration stage, which is defined by the constant state exactness and MPPT response speed requirements. The INC MPT algorithm normally uses a fixed phase of iteration based on the accuracy and the speed requirement for tracking.

The specific architecture will also properly resolve the equilibrium between dynamics and stabilization of oscillations, and the typical PV radiation is used to display the INC MPPT.

3 Proposed System and Control Strategy

The proposed system for grid interfaced PV system in Fig. 5 is controlled with conventional PI and PR controller. In this system, the PV is designed with INC MPPT technique and applied to voltage source inverter [3, 4, 8, 9], which is used to regulate the DC link voltages and grid currents. Proportional Resonant control technique is implemented to control the grid currents and to generate gate signals required for voltage source inverter. For steady-state analysis, the DC bus voltage V_{DC} and the grid voltage Vac are assumed to be steady without disturbance.

The grid-linked inverter builds two control loops: the internal control loop and the external control loop. The internal loop is faster and carries out the current injected quality and power between the grid and the system. In-loop referent current is generated by the voltage control circuit. The VSI is interconnected with [10] parameters as shown in Fig. 5 by means of an LCL filter. Filter reduces performance current distortion. For grid link, the grid norm must be achieved by inverter sinusoidal tension rating. It is necessary to have a passive filter. L is the quickest and most popular filter. The large meaning of L will, therefore, be added in order to rise the existing ripple. The expense and weight of the unit is also reduced by utilizing the RL filter. Active strategies for damping the grid-connected inverter due to filter resonances are nevertheless required to stabilize the device[15]. The added damping resistor in series L raises energy losses by the usage of a resistor that reduces unit performance. Damping should then be performed in C series, and the strength of

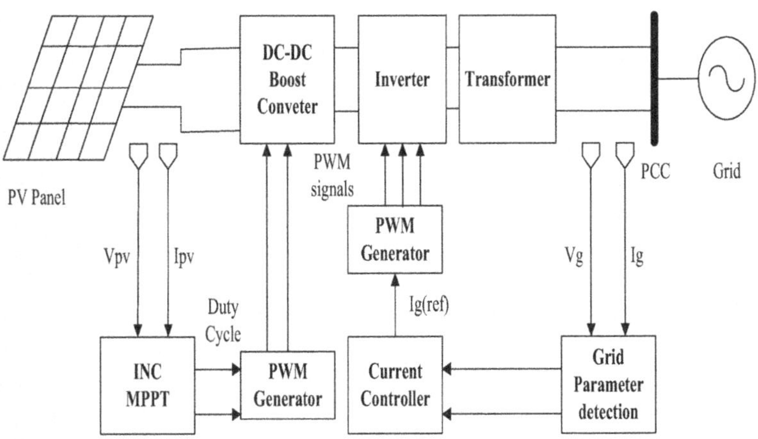

Fig. 5 Block diagram of the proposed PV grid system

the resistor can be reduced. Therefore, the grid is not always reliable because of the poor network climate. Net Voltage has several perturbations including scaling, scaling and frequency distortion. For robustness, the grid-connected quality system is needed. This article includes the configuration of a PV inverter with single-phase grid connections with PR controls. This grid-connected inverter is adapted to the RL converter. Simulations are conducted and the accurate performance is seen during several disruptions. This control model is utilized by multiple controllers such as (a) Proportional Integral Controller, (b) Proportional Resonant Controller to obtain a more precise interpretation of condition.

A contrast between [5–7] is provided and addressed. A general flaw in the PI controller is that without an integral term structure, it can't adopt a sinusoidal relationship. The inability to track a sinus connection activates the desire to use grid tension as a feed term to obtain a successful dynamic response, allowing the controller to attempt to achieve faster steadiness. The PR system is a modern system more suited for sinusoidal interactions that has no drawbacks. Around a certain frequency (resonant frequency), the PR controller produces gain and almost no improvement at other frequencies. This article describes the proposal for a single-phase, RL filter system, and current power, linked PV inverter. The PI-PR controllers used in grid-connected PV systems are often distinct with test findings in different cases being used.

3.1 Proportional Integral Control

PI controllers are commonly called as conventional controllers [6], which helps to minimize the steady-state errors of the regulators. In this, the value of the output signal is proportional to error signals with minimized steady-state error. Conventional PI controller may have high gain in the DC signal of a polar source, which is ideal for regulating the voltage range

The general concept is for the PI controller.

$$U_e = K_p e(t) + \int K_i e(t)dt \tag{1}$$

In this system, PI controller is applied at inverter control diagram to regulate the grid and DC link voltages. The PI controller used here is type-1 system which helps to minimize the steady-state error.

Because of its simplicity, the PI controller has been used for many years in many applications. The new PI controller has been designed for the THD reference. GPI(s) is known as the current PI controller

$$u(t) = K_p e(t) + K_i \int_0^t e(t)dt \tag{2}$$

The PI device was designed for a damping factor of 0.8 and a normal frequency of 3142 rad/ sec, which culminated in a K p of 4.21 and a KI of 2107. Application: The obtained damping factor was 0.85, while the natural frequency was 3360 rad per second.

3.2 Proportional Resonant Control

The PR control unit, but in terms of length, stage, and range, incorporates PR frequencies past resonance frequency in the integer creating a steady error in the state and phase shifting[16]. The PR control device is thus identically configured to the PI system. The harmonic order that can be regulated without adjustments in the stability limits is also calculated. The PR regulator produces a very strong response boost in a low-frequency band with resonant frequency gains that do not impact the sweep of the PR controller because it only recognize frequency very similar to resonance oscillation[11]. The sensitive frequency compensator may be regulated by applying a couple of gains to the resonance at the desired frequency. The PR methodology should also be used extensively in the existing grid converter monitoring applications where the existing needs to be matched to the fixed grid frequency and the weak harmonics are compensated[5]. The proportional resonant control system is the best way to better track a reference and eliminate the interrupted signal as it focuses on the internal model [7].

The mathematical model representation for PR controller is

$$G_{\mathrm{PR}}(s) = K_P + \frac{2K_r s}{s^2 + w_o^2} \tag{3}$$

The system stability with an above-mentioned mathematical analysis is achieved using an internal model main and ensures that no constant state error can be achieved with sinusoidal system frequency tracking.

An approximating (nonideal) PR system is implemented to prevent reliability issues synonymous with infinite gain.

A low-pass high-gain filter is used when a nonideal PI controller can be replaced by the nonideal PR controller.

$$G_{PR}(s) = K_P + K_i \frac{(\omega_c s + \omega_c^2)}{s^2 + 2\omega_c s + \omega_c^2 + w_o^2} \tag{4}$$

It understands where KP and Ki are so the frequency of resonation is similar. At the real frequency AC, the PR method generates an infinite increase. However, regardless of losslessness for the ideal resonant PR device, no phase change is feasible. A perfect PR controller also allows for infinite advantage in stability problems. It allows the PR system perfectly to escape these difficulties. The nonideal PR control gains may be modified such that the voltage monitoring error is reduced by obtaining ample end gain. The PR controller gains can be built into a step-by-step process.

At the fundamental frequency, the PI controller has a steady benefit. Kbe does not ignore the second word, but it can be excluded by the decoupling technique. Nonetheless, the PR control mechanism involves a constant boost, and then the first term hits the true partnership between the inverters.

$$H_{iR}(s) = \frac{A_3s^3 + A_2s^2 + A_1s + A_0}{B_5s^5 + B_4s^4 + B_3s^3 + B_2s^2 + B_1s + B_0} \tag{5}$$

$$A_3 = R_dC_fK_pK_i \tag{6}$$

$$A_2 = K\left[K_p + \left(2K_p + K_i\right)\omega_cR_dC_f\right] \tag{7}$$

$$A_1 = K\left[\left(2K_p + K_i\right)\omega_c + K_pR_dC_f\omega_0^2\right] \tag{8}$$

$$A_o = KK_p\omega_o^2 \tag{9}$$

where an infinite rise at the fundamental frequency is implemented by the PR transmitter. The inverter would have a marginal effect on the dynamic control loop, and the input voltage is called sinusoid, the real closed-loop control stability can be measured by selecting the VSI switching frequency higher than the device frequency. The PR device is built for 0.5 rad/ s, achieving a KP of 5.1 and KI of 2073.15. The resonant strength is up to < 0 of 314.2 rad/s (50 Hz). Using the PI controller would be a problem for the sinusoidal current tracking control in a stationary frame because the gain in grid frequency is limited. A proportional resonant (PR) internal-scale controller with high gain at the grid frequency will simplify the control-loop configuration and increase device output in order to maintain a close sinusoidal current monitoring control for the grid-tied inverter. The literature has reported several successful applications for the PR controller [7]. A powerful PR system was incorporated to help take the grid impedance into account. The usage of a pic controller will have a problem with sinusoidal current tracking of −β axes for the use of the current decoupling control because the grid frequency control gain is small. A controller with high gain at the grid frequency will promote control-loop configuration and enhance device efficiency for close sinusoidal current monitoring.

4 Simulation Case Study

In order to validate the proposed system, this Simulink model is tested and verified in MATLAB environment under different cases. And the simulation results for all cases are shown below where active and reactive powers are given with different values

Table 1 System parameters

Parameter	Value/Description
Grid Voltage (Vg)	120 V (R.M.S)
DC Link Voltage	48 V
Switching Frequency	5 kHz
Fundamental Frequency	50 Hz
Single-Phase Transformer	21 V/120 V
Inverter Topology	1-ϕ Full Bridge Inverter
PWM Technique	Unipolar Sine Triangle
Filter	L type

Table 2 Simulation results of active and reactive power compensations with PI

Testing conditions	Reference Input Active power P(ref)	Reference Input Reactive power Q(ref)	Grid output Active power P with PI	Grid output Reactive power Q with PI	Grid output Active power P with PR	Grid output Reactive power Q with PR
Case-I	70	20	71.22	20.88	70.1	20.04
Case-II	30	60	31.78	52.02	30.07	60.1
Case-III	60	60	61.19	57.48	60.11	60.1

under rated and above rated, Simulation results of Proportional Resonant Controller, and PI controller. The system parameters are shown in Table 1, according to this details model designed in Simulink. the simulation results of PI and PR controllers for active and reactive power compensations are shown in Table 2.

System Descriptions:

The simulation results of active and reactive powers of PV system with different compensation powers and those comparisons are noted in a table as we compared all results of PR controller give better performance more than PI controller. When the power compensation was done with active power 60 W and reactive power with 0VAR thrn the single-phase PV grid system under constraints of without PI controller, PI controller and PR controller the steady error shows a value of 13%, 7% and very less as comparing steady-state errors also PR controller gives best performance. As PR controller having numerator coefficients of [1.202 2000 1.709e5] and Denominator coefficients of [1 0 1.421e5], where Kp and Ki values are 1 and 0.05094 for effective understanding of active and reactive power compensations the model is running at three different modes.

Mode I: Without Controller

In this mode, the designed system performed without controller, here compensation is not obtainable as shown in Fig. 6.

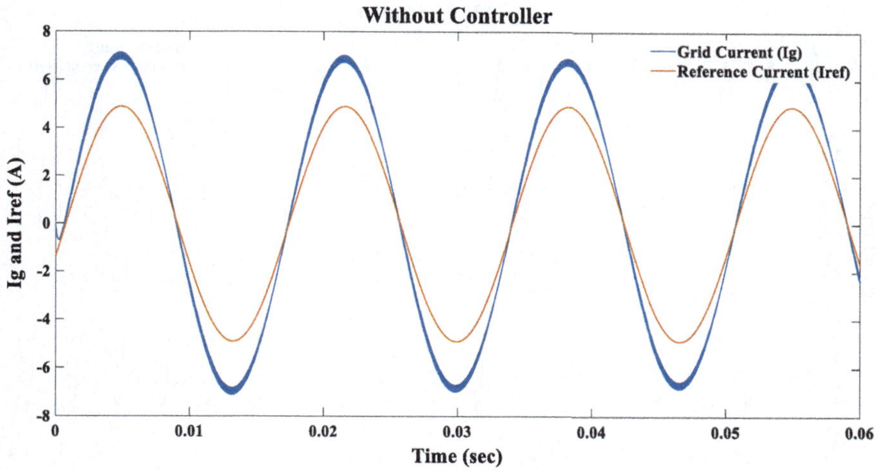

Fig. 6 Compensation without controller

This waveform clearly shows the actual grid current and reference grid current compensation is not occurred by using without controller.

Mode II: With PI Controller

In this mode, the designed PV system was performed with PI controller, here compensation is obtained from 0.228 s as shown in Fig. 7.

Initially, 70 W active power and 20VAR reactive powers are given as a reference to observe the results of the system with PI and PR controllers.

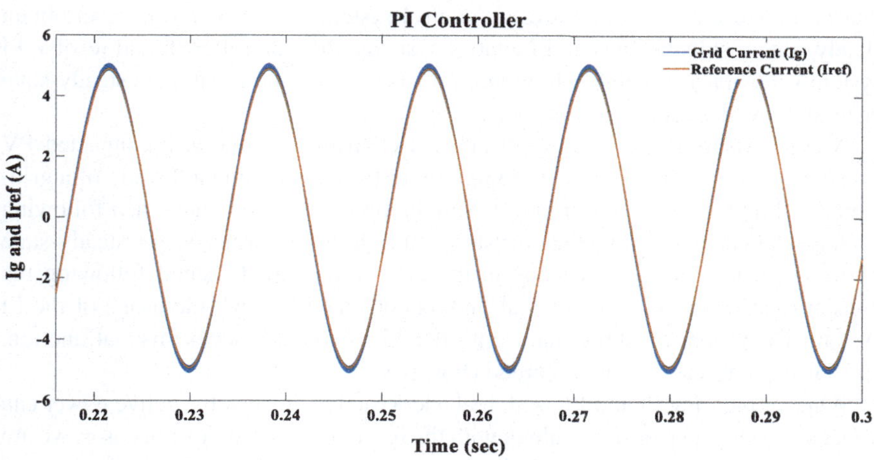

Fig. 7 Compensation with PI controller

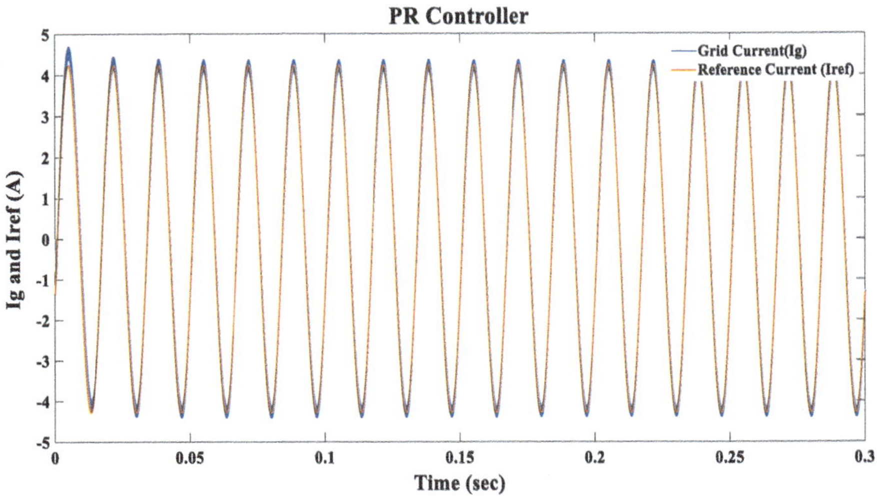

Fig. 8 Compensation with PR controller

Mode III: With PR Controller

In this mode, the designed PV system is performed with PR controller, here compensation is perfectly obtained immediately and its value is nearly less than 0.001 s, which was shown in Fig. 8, by observing this figure compensation of grid current and reference current gives a better result of active and reactive power compensation at grid side. All performed results are shown in the table where the multiplied power in sim block maintains the same power as compensation is doing well. In accordance with all of the above- described experimental findings, it can be seen that in the current-controlled inverter control system, the PR controller can obtain steady-state results better than PI control systems. PR Controller should resolve PI system's drawbacks: inability to monitor a sinew connection with zero steady-state error and low destructive disruption.

A comparison of the existing PI and PR controllers in the grid-connected PV Inverter was given. This article's simulation tests and experimental study of a grid-connected inverter are seen in Fig. 9. Simulation results show that when following a sinusoidal reference, a voltage forward controller is suffering from a steady-state error. There was an inconsistency in the grid current fundamental following the present relation in the experimental findings obtained through the usage of the PI system. The proportional resonant controller efficiently eliminates this malfunction, and subsequent readings are recorded (Fig. 10).

When comparing PI and PR with real and Reactive Powers, the active power can affect the system performance along with the reactive power in most of cases, we are taking results of PV grid system with high active power and low reactive power, later on increasing reactive power in the same with or more than active power. This PV Grid system is compared in three cases, in the first case without control the reference

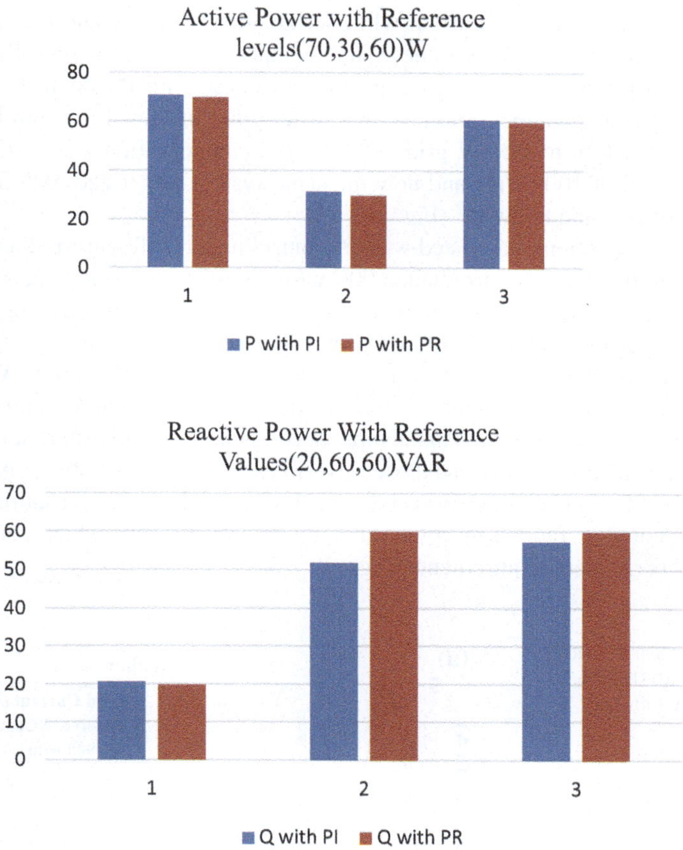

Fig. 9 Comparative Results of PI and PR controller for active and reactive power compensation

Fig. 10 Comparison PI and PR with different reference values of active power and reactive power

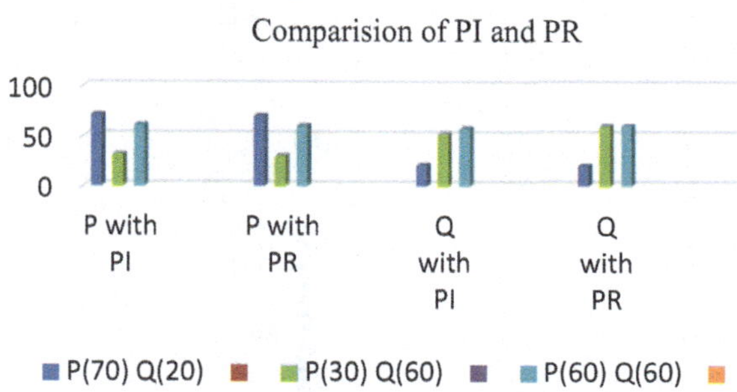

grid current and actual grid current will overlap after 9th cycle and it is not given proper compensation when the same system compensated with PI controller it takes 5 cycles to overlap each other, by checking third case with PR controller it takes 1cycle for effective compensation as final result point of view PR controller gives faster compensation to the PV grid system. This compensation was tested with a irradiance level of 1000w/m^2 and now the same system was tested with irradiance level of 900w/m^2 and 800w/m^2 (Fig. 11).

The PV Grid system was tested with PI control under different irradiance levels the compensation time for irradiance 900 w/m^2 is 0.23 and irradiance 800 w/m^2 is 0.242, respectively. In this comparison of PI controller performed for different irradiance levels of 1000 w/m^2, 900w/m^2, and 800w/m^2. Along with these results, PI and PR controllers with different irradiance levels were performed in MATLAB to analyze system response with time by utilizing internal loop which compares the grid current and reference current. This comparison signals generate different responses for different irradiance levels, and those performance are noted in Table 3. According to this table, PR controller performs better as in the previous compensation method which was noted in Table 2 by comparing all simulation results which controller is suitable as per system requirements (Figs. 12, 13, 14, 15 and 16).

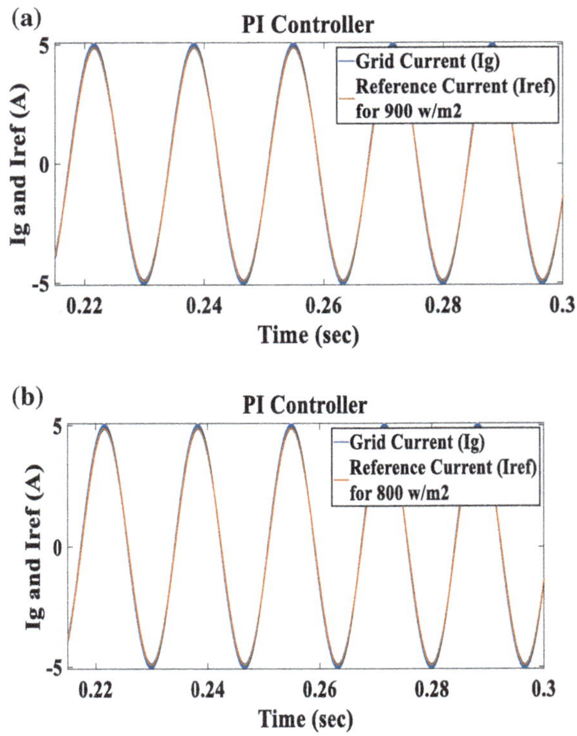

Fig. 11 Comparison of PI controller with shading conditions **a** 900w/m^2 **b** 800w/m^2

Table 3 Simulation results of PI and PR controllers with different irradiance values

Irradiance (w/m^2)	Compensation time with PI in sec	Compensation time with PR in sec
Irradiance ($1000w/m^2$)	0.228	0.01
Irradiance ($900w/m^2$)	0.23	0.018
Irradiance ($800w/m^2$)	0.242	0.022

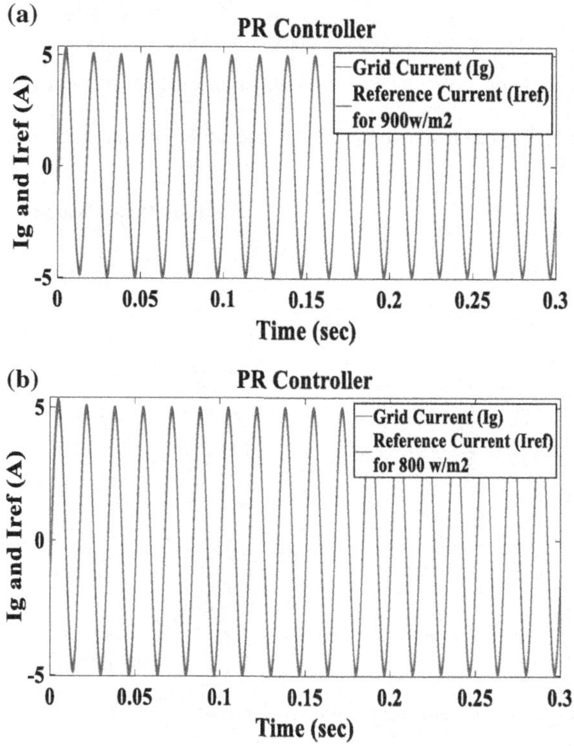

Fig. 12 Comparison of PR controller with shading conditions **a** $900w/m^2$ **b** $800w/m^2$

In all cases, the PV panel output voltage for three irradiations (800, 900, and1000) w/m^2 are noted as 17.8, 17.85, and 17.9 Voltages. At that irradiation level, the output voltage of boost converters are 75.9, 76.2, and 76.78 Volts.

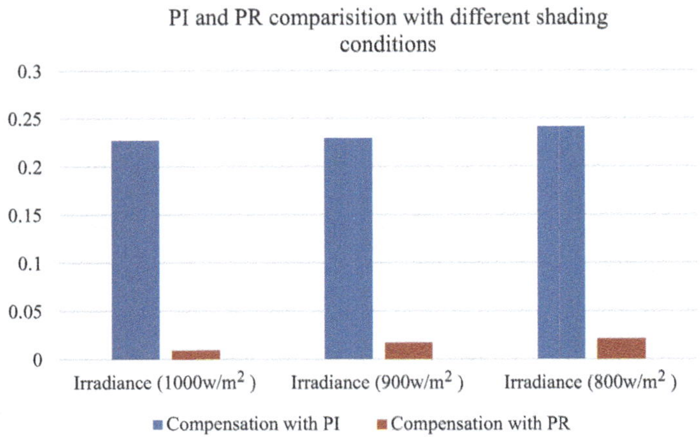

Fig. 13 Comparative Results of PI and PR controller with shading conditions

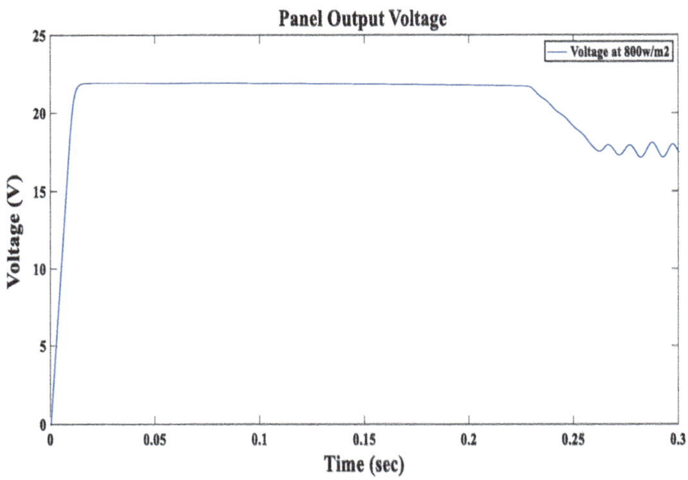

Fig. 14 PV Panel output voltage for irradiance of $800w/m^2$

5 Conclusion

This paper has applied reactive power management for grid-connected PV network with different control strategies. The suggested simpler reactive energy controller will impact on the processor measurement such that an inexpensive PV inverter converter can be applied. To show power stability, a limited signal analysis is carried out for the PR system. When evaluating both the effects of different active power and reactive power compensations, the most efficient device is the PR device, which can be used to achieve the optimal active, reactive power regulation over the proposed

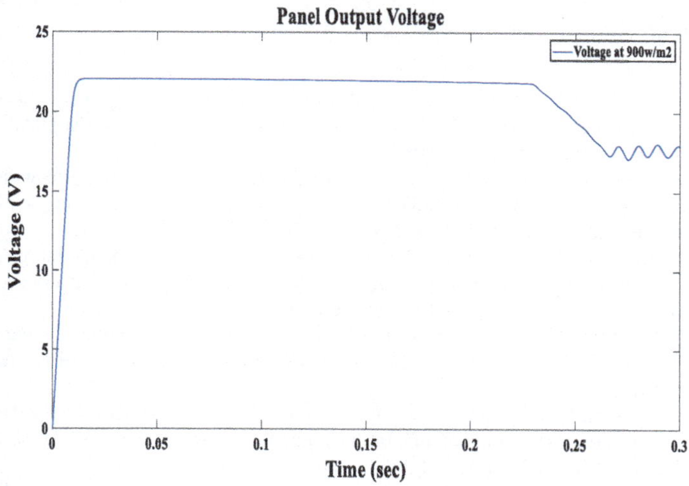

Fig. 15 PV Panel output voltage for irradiance of 900w/m^2

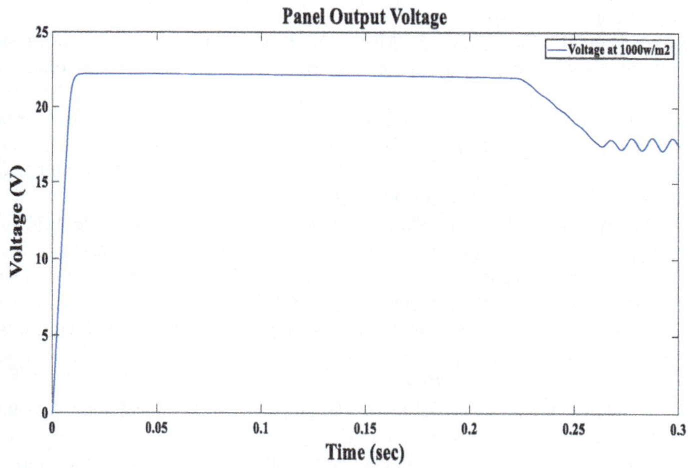

Fig. 16 PV Panel output voltage for irradiance of 1000w/m^2

single-phase grid-connected PV network. After comparing with the PI system, the feedback from the PR controller is easily used to enhance the output.

- Active and reactive power compensation done effectively with PR controller as compared with PI controller.
- Different shading conditions applied on the system by using PI and PR controllers, in this point also PR controller gives fast response.

References

1. H. Kanchev, D. Lu, F. Colas, V. Lazarov, B. Francois, Energy management and operational planning of a microgrid with a PV-based active generator for smart grid applications. IEEE Trans. Ind. Electron. **58**(10), 4583–4592 (2011)
2. G.G. Talapur, H.M. Suryawanshi, L. Xu, A.B. Shitole, A reliable microgrid with seamless transition between grid connected and islanded mode for residential community with enhanced power quality. IEEE Trans. Ind. Appl. **54**(5) (2018)
3. Q. Tabart, I. Vechiu, A. Etxeberria, S. Bacha, Hybrid energy storage system microgrids integration for power quality improvement using four-leg three-level npc inverter and second-order sliding mode control. IEEE Trans. Ind. Electron. **65**(1) (2018)
4. M. Adly, K. Strunz, Efficient digital control for MPP tracking and output voltage regulation of partially shaded PV modules in DC bus and DC microgrid systems. IEEE Trans. Power Electron. **34**(7) (2019)
5. R. karthik, A. Miruthula, N. Nitheeswari, Web based online machine controlling and monitoring using PLC via modbus communication. Int. J. Commun. Comput. Technol. **7**(2), 22–26 (2019). https://doi.org/10.31838/ijccts/07.02.06
6. M.J. Naik, mapk signalling pathway: role in cancer pathogenesis. J. Crit. Rev. **6**(3), 1–6 (2019). https://doi.org/10.22159/jcr.2019v6i3.31778
7. H.P.H. Anh; L. Vinh Truong, C. Van Kien, Advanced intelligent fuzzy control of standalone PV-wind-diesel hybrid system, *2019 International Conference on System Science and Engineering (ICSSE),* 05 Sept 2019 (2019)
8. W. Sinsukthavorn, E. Ortjohann, A. Mohd, N. Hamsic, D. Morton, Control strategy for three/four wire-inverter-based distributed generation. IEEE Trans. Ind. Electron. **59**(10), 3890–3899 (2012)
9. R.P. Agus, Design of reliable and efficient manchester carry chain adder based 8-Bit Alu for high speed applications. J. VLSI Circ. Syst. **1**(01), 1–4 (2019)
10. M.I Ahmed, A compact triangular ring patch antenna for radio location and fixed satellite applications. Nat. J. Antenna. Propag. **1**(1) (2019)
11. S. Sheik Mohammed, D. Devaraj, Simulation of incremental conductance MPPT based two phase interleaved boost converter using MATLAB simulink, in *IEEE* (2015)
12. T.K. Vu, S.J. Seong, Comparison of PI and PR controller based current control schemes for single-phase grid-connected PV inverter 11(8), 2968–2974 (2010)
13. H. Patel, M. Gupta, A.K. Bohre, Mathematical modeling and performance analysis of MPPT based solar PV system, in *IEEE International Conference on Electrical Power and Energy Systems (ICEPES)*, pp. 157–162 (2016)
14. S. Rhili, H. Trabelsi, J. Hmad, Modeling of a singlephase grid-connected photovoltaic system. STA 563–567 (2017)
15. H. Kancherla, S. Syama, Simulation and analysis of incremental conductance and perturb and observe MPPT with DC-DC converter topology for PV array, in *2015 IEEE International Conference on Electrical, Computer and Communication Technologies (ICECCT)* (IEEE, 2015)
16. J. Hmad, H. Trabelsi, S. Rhili, ROCOF anti-islanding strategy based on frequency tracking observer for three phase DG inverter (2018)
17. R.N. Bharti, R.K. Mandal, Modeling and simulation of maximum power point tracking for solar PV system using perturb and observe algorithm. Int. J. Eng. Res. Technol. **3**(7), 7–18 (2014)
18. V.K. Viswambaran, A. Ghani, E. Zhou, Modelling and simulation of maximum power point tracking algorithms & review of MPPT techniques for PV applications, in *2016 5th IEEE International Conference on Electronic Devices, Systems and Applications (ICEDSA)*, pp. 1–4 (2016)
19. C. Mihai, T. Remus, B. Frede, a new single-phase PLL structure based on second order generalized integrator, in *IEEE Annual Power Electronics Specialists Conference PESC Record* (2016)

Design and Analysis of Fifth-Order Bi-Directional Charger with Vehicle to Grid Application

Ashish Kumar Singhal, Narendra S. Beniwal, and Rajesh Kumar

1 Introduction

1.1 Background

Bidirectional charging can be possible with DC-DC converter having bidirectional capability, which allows the flow of power in both directions. These converters are especially suited for applications in traction because of their conversion properties [1, 2]. Bidirectional charging with (V2G) application is the most delirious concept for the electric vehicle (EV) charging industry. The (V2G) technology permits us to communicate with the energy system to stabilize the power grid through demand response [3].

With an unassuming word, bidirectional charging with V2G allows us to not only charge the batteries of EV but also take energy from car batteries and push it back to the power grid to help balance momentary spikes in electricity demand [4].

Indeed! the power grid and the globe, both will be benefited from V2G, which makes it possible to reserve renewable energy and pull it back into use [5]. Without V2G, ample amount of renewable energy would go to waste (Fig. 1).

A. K. Singhal (✉)
Sagar Institute of Science, Technology and Engineering, Bhopal, India
e-mail: ashishee@gmail.com

N. S. Beniwal
Bundelkhand Institute of Engineering and Technology, Jhansi, India
e-mail: narendra.beniwal@gmail.com

R. Kumar
MNIT, Jaipur, India
e-mail: rkumar.ee@mnit.ac.in

V. K. Singh et al. (eds.), *Renewable Energy and Future Power Systems*, Energy Systems in Electrical Engineering, https://doi.org/10.1007/978-981-33-6753-1_5

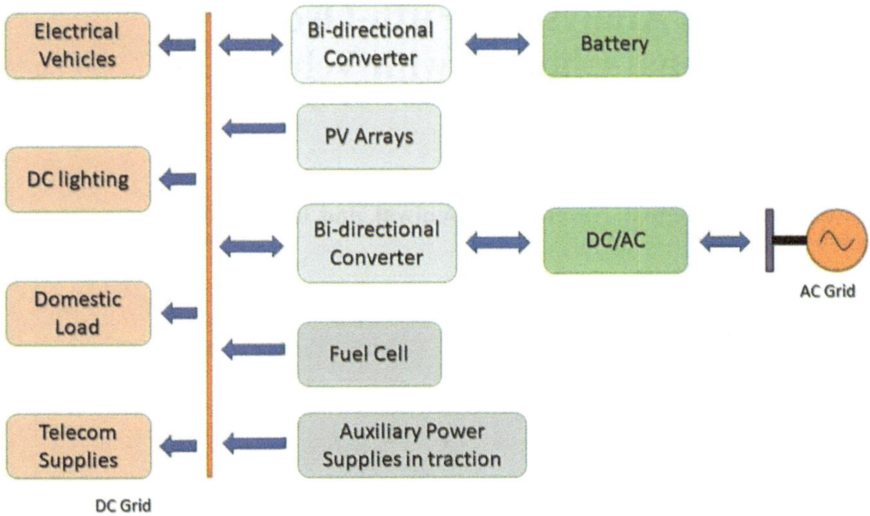

Fig. 1 Bidirectional DC-DC converter in micro-grid [2]

1.2 Bidirectional DC-DC Converter

Bidirectional Converter permits power flow in either directions without change of load [5]. This converter's supply and storage voltage levels will be different in either directions. So, Converter operates in buck and boost mode for different directions either forward or reverse. In the literature, there were certain problems of high input (I/P) or output (O/P) current as well as voltage ripples. The conversion range of these converters are limited. To achieve stable performance, these converters need to have large number of switches with good voltage conversion range [6]. This configuration gives low-current ripple and low-voltage ripple with three switches. Due to presence of the inductors at both (I/P) and (O/P) side of the converter, the current ripple is reduced at either end. Bidirectional converters have a wide range of applications [4, 5] in battery charger circuits such as telecom, computer power systems, automotive, etc. They have major applications in DC micro-grids [5] for load leveling.

1.3 Performance Indices

To assess the predominance of a converter topology, we have to consider the factors that are affecting performance of bidirectional converters.

1.3.1 Source and Load Current Ripples

The life of a battery depends on the current ripple in charging current [7]. High-ripple current causes internal corrosion of batteries and temperature rises. So, to improve the battery life, there is a need to realize lower ripple in battery charging current. But, always increasing the inductance for minimizing the ripple is not a solution. As soon as the size of converter increases, the Electro-Magnetic Interference problems become predominant [8].

1.3.2 Switch Voltage and Current Stresses

For low-voltage stress in the circuit, the rating of switches will be lowered [7] which will reduce the cost of converter. Beyond this, some converters may suffer from capacitor voltage balancing problems due to which the converter's performance will deteriorate. The higher current stresses will create local hot spots in switches, reducing their life. In IGBTs, the peak currents during switching causes extended tail currents which limits the operating frequency range.

1.3.3 Conversion Range

Higher conversion range is a desirable characteristic for any DC-DC converter. If we can get higher bucking or boosting action at lower duty ratios, then it can be minimized on-time of switches which ultimately reduces the conduction losses [9].

1.4 Review of Bidirectional Converter Topologies

1.4.1 Conventional Bidirectional Converter

Conventional Bidirectional converter topology is replacing the shunt diode of buck converter with a switch [5]. It is the simplest bidirectional converter configuration with just two switches with passive element in Fig. 2.

During charging operation, switch S_2 will perform under PWM switching. Figure 2 shows buck converter topology, other switch driver circuits will not be triggered whereas its anti-parallel diode helps in circulating current on the load side. During discharging operation, only switch S_1 will undergo PWM switching while switch S_2 driver circuit will not be triggered. The configuration works as boost converter topology. Anti-parallel diode of switch S_2 will provide path for power flow to high-voltage side. The conversion gain for this converter reported as follows.

$$\text{Charging Operation } V_{LV} = V_{HV} D \tag{1}$$

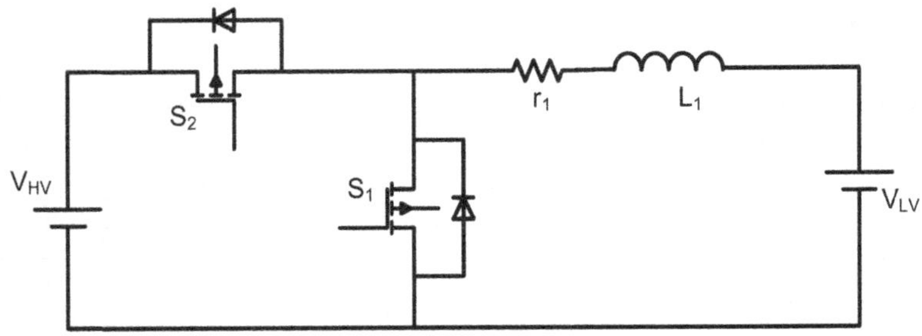

Fig. 2 Conventional Bidirectional DC-DC Converter

$$\text{Discharging Operation } V_{HV} = V_{LV}\left(\frac{1}{1-D}\right) \quad (2)$$

Conventional Bidirectional converter has a very simple topology to realize but suffers from high voltage and current stresses, along with the drawback of limited conversion range [7].

1.4.2 Third-Order Buck–Boost Converter

Third-order Buck–Boost Converter was developed to overcome the current stresses experienced by the conventional Bidirectional Converter [7, 10]. The problem of high current stress will be reduced to some extent due to the presence of inductor in the circuit (Fig. 3).

For charger operation only switch S_2 will undergo pulse width modulated switching. During turn-on state, the current flows from source through inductor L_1

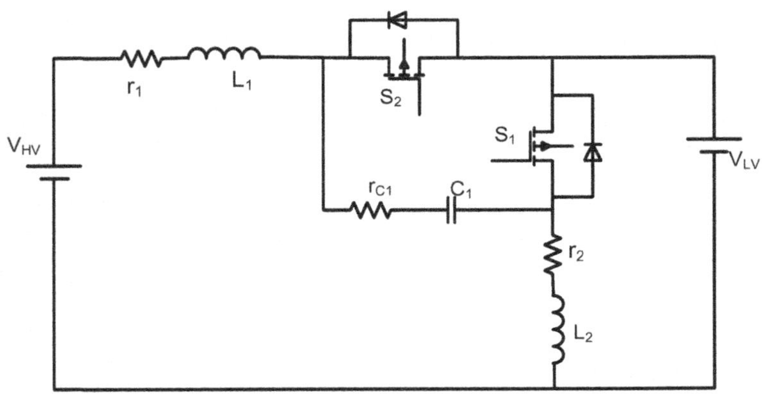

Fig. 3 Third-order Bidirectional DC-DC Converter

and switch S_2 to load. When switch S_2 is turned-off, anti-parallel diode of switch S_1 helps in circulating current on the load side.

During discharger operation, only switch S_1 will undergo pulse width modulated switching while switch S_2 driver circuit will not be triggered. When switch S_1 is turned-on, inductor L_2 will store energy. When switch S_1 is turned-off, inductor L_2 along with the source pumps energy through anti-parallel diode of S_2 and inductor L_1 to high voltage side. The conversion gain for this converter is reported as follows.

$$\text{Charger Operation } V_{LV} = V_{HV} D \tag{3}$$

$$\text{Discharging Operation } V_{HV} = V_{LV}\left(\frac{1}{1-D}\right) \tag{4}$$

Even though the number of energy storage elements was more than the conventional bidirectional converter, it means the order of converter increases, still the conversion range was the same as the conventional bidirectional converter.

1.4.3 Wide Voltage Conversion Range Converter

In this topology, driver circuits for all switches have to undergo pulse width modulated switching whether in charger or discharger operation [9]. So voltage stress is less on the switches compared to other converters but on-state conduction losses in switches will account for the reduction in efficiency. The conversion gain for this converter is reported as follows (Fig. 4).

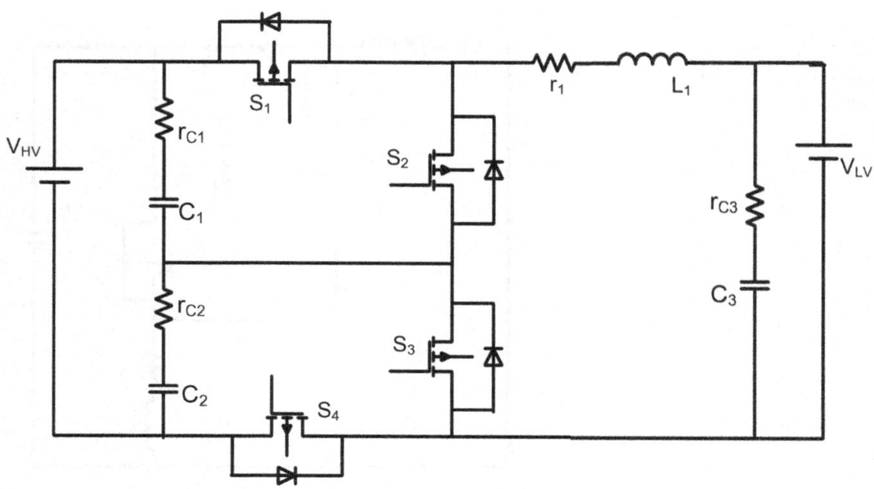

Fig. 4 Wide voltage conversion range bidirectional DC-DC Converter

$$\text{Charger Operation } V_{LV} = V_{HV}\left(\frac{D}{2}\right) \tag{5}$$

$$\text{Discharger Operation } V_{HV} = V_{LV}\left(\frac{2}{1-D}\right) \tag{6}$$

In this converter, the conversion range is improved and the current stresses are reduced to a greater extent. To realize this performance, it needs four switches also separate driver circuits were required which increase the cost of the circuit. High-voltage side capacitors will have great effect on the converter performance. In this configuration, we have to maintain the C_1 and C_2 as the same value. But temperature will bring deviation from the original value. The unequal capacitor values create disturbance in voltage balance. This will increase voltage stresses in the circuit. Also, if we have to rectify this problem, capacitor charge balance methods have to be employed [11]. These will increase the cost and complexity of the circuit.

2 Steady-State and Dynamic Analysis of the Proposed Bidirectional Charger

The proposed topology is having three switches. As shown in Fig. 5, for the proposed converter the switches are represented with S_1, S_2, and S_3, whereas diodes by D_1, D_2, and D_3. Based on power flow direction, the switching will change and the switching operation for charger and discharger operation is shown in Table 1. The working of the converter in each mode is described in the following sections.

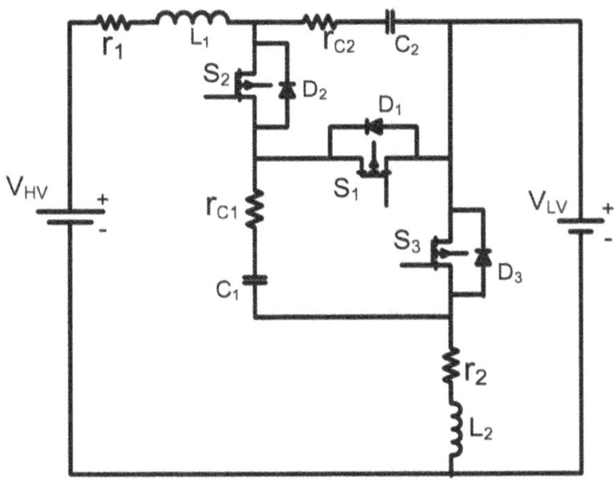

Fig. 5 Proposed Bidirectional DC-DC converter

Table 1 Status of switching devices for charger operation

Operation	Charger		Discharger	
	$0<t<DT_s$	$DT_s<t<T_s$	$0<t<DT_s$	$DT_s<t<T_s$
Switches	S_1	S_2	–	S_3
Diodes	–	D_3	D_1	D_2

2.1 *Mathematical Analysis of Charger*

The power flow from DC grid to battery, i.e., charging the battery is called as charger operation. In charger operation, switches S_1 and S_2 will undergo pulse width modulated switching. Switch S_3will remain in turned-off during charger operation. When switch $S_1$is turned-on, S_2 and D_3 are turned-off and the charger will operate in mode-1, when S_1 is turned-off, S_2 and D_3 will conduct, then circuit enters into mode-2.

Mode-1 ($0 < t<DT_s$)

Switch S_1 is turned- on while switch S_2 remains in turned-off state. DC grid supply charges the battery through inductor L_1. This mode terminates with switching off of S_1 switch (Fig. 6).

Mode-2 ($DT_s< t<T_s$)

In this mode, S_1 switch is turned-off while switch S_2 and D_3 are going to conduct. In this mode both the capacitors are coming in parallel. The source side circuit V_{HV} and inductor L_1 are charging capacitors $C_1$and C_2. This mode terminates with turn-on of switch S_1 (Fig. 7).

Conversion Gain During Charging

The voltage gain of the converter is derived from volt-sec balance of inductors in all modes. The converter has two modes of operation. For simplicity, the non-idealities are neglected and the load-side battery has been replaced with parallel *R* and *C* combination, and further analysis will follow the same convention (Fig. 8).

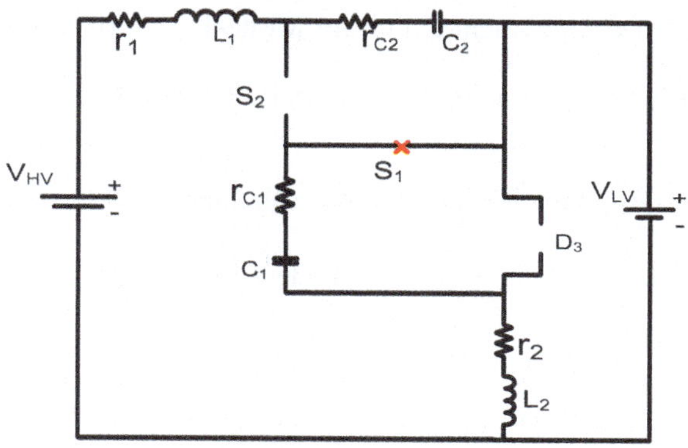

Fig. 6 Equivalent Circuit of charger during mode-1

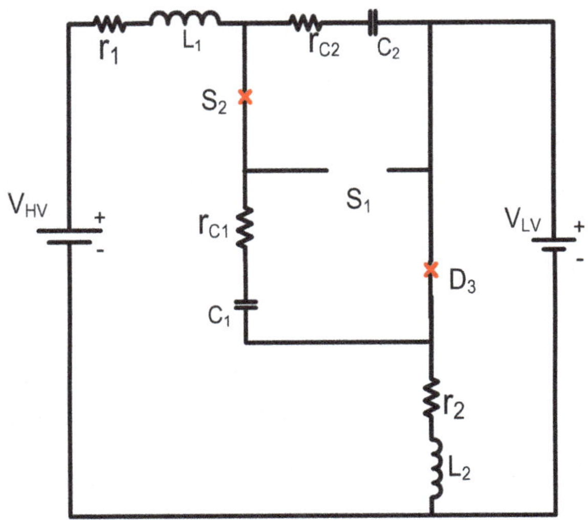

Fig. 7 Equivalent circuit of charger during mode-2

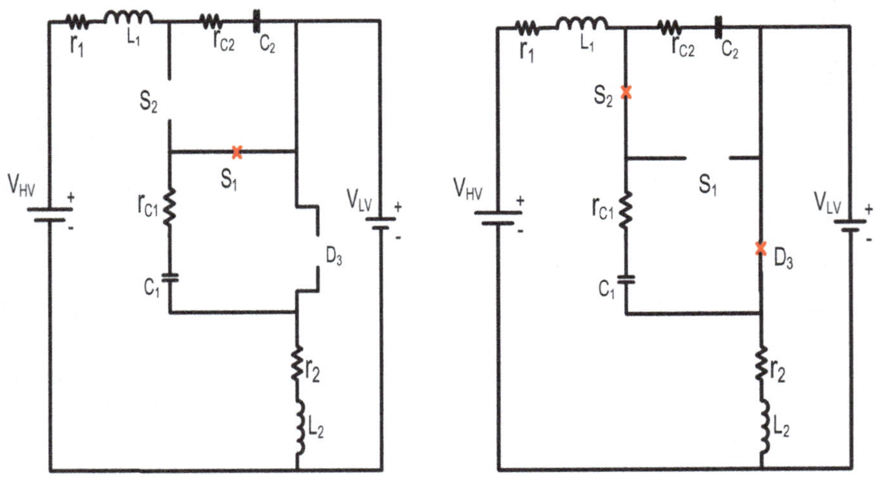

(a) Equivalent circuit for Mode-1 (b) Equivalent circuit for Mode-2

Fig. 8 Equivalent circuits of charger in CCM mode

Applying KVL in the above circuits,

$$\text{For mode-1} \quad v_{L1} = v_{HV} - v_{C2} - v_{LV} \tag{7}$$

$$\text{For mode-2} \quad v_{L1} = v_{HV} - v_{C2} - v_{LV} \tag{8}$$

Applying volt-sec balance for inductor L_2 and on simplification results in

$$(v_{LV} - v_{C1})D + v_{LV}(1 - D) = 0 \tag{9}$$

$$v_{LV} - v_{C1}(D) = 0 \tag{10}$$

$$v_{C1} = \frac{v_{LV}}{D} \tag{11}$$

Applying volt-sec balance for inductor L_1 and on simplification results in

$$(v_{HV} - v_{C2} - v_{LV})D + (v_{HV} - v_{C2} - v_{LV})(1 - D) = 0 \tag{12}$$

$$v_{HV} - v_{C2} - v_{LV} = 0 \tag{13}$$

$$v_{HV} = \frac{v_{LV}}{D} + v_{LV} \tag{14}$$

$$v_{HV} = v_{LV}\left(\frac{1 + D}{D}\right) \tag{15}$$

$$V_{LV} = V_{HV}\left(\frac{D}{1 + D}\right) \tag{16}$$

2.2 *Mathematical Analysis of Discharger*

The power flow from battery to DC grid called as discharger operation. In discharger operation only switch S_3 will undergo pulse width modulated switching. The switches S_1 and S_2 will remain turned-off during discharger operation. When switch D_1 is turned-on discharger will operate in mode-1, when D_1 turned-off and D_2, S_3 will start conducting, thus circuit will enter into mode-2.

Mode-1 ($0 < t < DT_s$)

In this mode, only diode D_1 is turned-on while D_2 and S_3 remain turned-off. The inductor L_2 is charging from battery similar to simple boost topology while the energy stored in L_1, C_1 and C_2 supplied to DC grid on the other part of circuit. This mode terminates when switch D_1 is turned-off (Fig. 9).

Mode-2 ($DT_s < t < T_s$)

Diode D_1 is turned-off in this mode. So, the anti-parallel diode D_2 and Switch S_3 are going to conduct allowing power flow from battery to grid side. The energy stored in

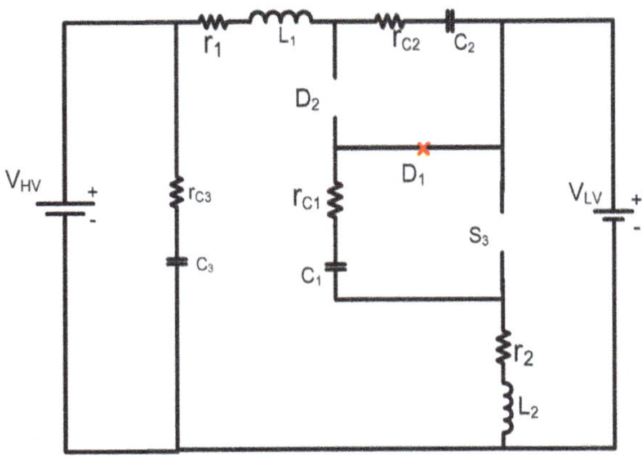

Fig. 9 Circuit Diagram of discharger during mode-1

inductor L_2 during mode-1 is now supplied to DC grid along with the battery. This mode terminates with turn-on of diode D_1 (Fig. 10).

Conversion Gain During Discharger

The voltage gain of the converter is derived from volt-sec balance of inductors in all modes. The converter has two modes of operation. For simplicity, the non-idealities are neglected and the load-side DC bus has been replaced with parallel R and C combination and further analysis will follow the same convention (Fig. 11).

$$\text{From MODE-1} \quad v_{LV} = v_{C1} + v_{L2} \tag{17}$$

$$v_{L2} = v_{LV} - v_{C1} \tag{18}$$

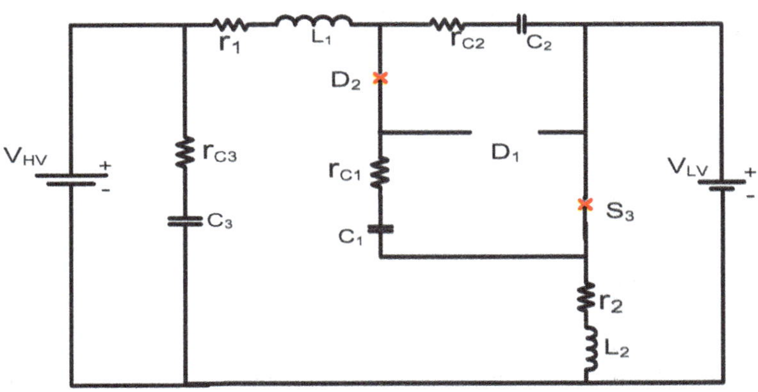

Fig. 10 Circuit Diagram of discharger during mode-2

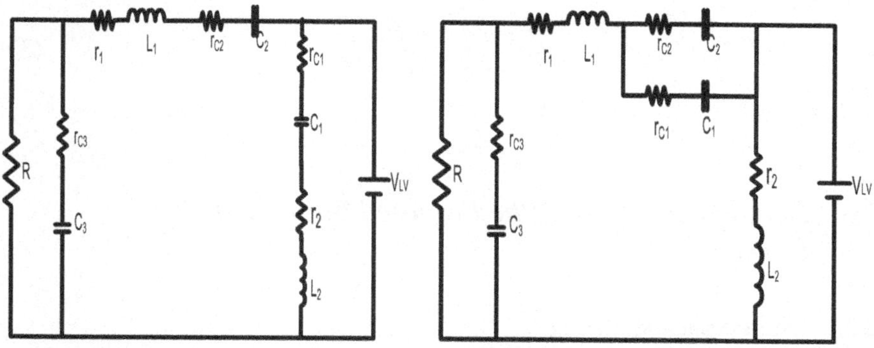

(a) Equivalent circuit for Mode-1 (b) Equivalent circuit for Mode-2

Fig. 11 Equivalent Circuits of discharger in CCM mode

$$\text{From MODE-2} \quad v_{L2} = v_{LV} \tag{19}$$

Applying volt-sec balance to inductor L_2 and on simplification results in

$$[v_{LV} - v_{C1}]D + [v_{LV}](1 - D) = 0 \tag{20}$$

$$v_{LV} - v_{C1}(D) = 0 \tag{21}$$

$$v_{C1} = \frac{v_{LV}}{D} \tag{22}$$

$$\text{From MODE-1} \quad v_{LV} = v_{C2} + v_{L1} + v_{HV} \tag{23}$$

$$v_{L1} = v_{LV} - v_{C2} - v_{HV} \tag{24}$$

$$\text{From MODE-2} \quad v_{LV} = v_{C2} + v_{L1} + v_{HV} \tag{25}$$

$$v_{L1} = v_{LV} - v_{C2} - v_{HV} \tag{26}$$

Applying volt-sec balance to inductor L_1 and on simplification results in

$$[v_{LV} + v_{C2} - v_{HV}](D) + [v_{LV} + v_{C2} - v_{HV}](1 - D) = 0 \tag{27}$$

$$v_{LV} + v_{C2} - v_{HV} = 0 \tag{28}$$

$$v_{LV} + \frac{v_{LV}}{D} = v_{HV} \tag{29}$$

$$v_{HV} = \left(\frac{1+D}{D}\right) v_{LV} \tag{30}$$

2.3 *Dynamic Analysis of the Proposed Bidirectional Converter*

2.3.1 State-Space Modeling

State-space modeling is the most useful tool in analyzing SMPS circuits for both CCM and DCM operations [12]. It is an analytical modeling of state-space equations portraying linear equivalent circuits for different states of the network by ON and OFF states of various switching devices in the circuit. The major advantage is it can include non-idealities of circuit elements in the analysis. So in this work state-space modeling technique is used for analyzing the proposed converter circuit in CCM operation by considering the non-idealities of components. [x] is state vector having state variables of the converter.

$$[x] = [\, i_1 \; i_2 \; v_{C1} \; v_{C2} \; v_{C3} \,]^T \tag{31}$$

$$[\hat{x}] = [\, \tfrac{di_1}{dt} \; \tfrac{di_2}{dt} \; \tfrac{dv_{C1}}{dt} \; \tfrac{dv_{C2}}{dt} \; \tfrac{dv_{C3}}{dt} \,]^T \tag{32}$$

[u] is input vector and [y] is the output vector. The state-space average model

$$[\hat{x}] = [A][x] + [B][u] \tag{33}$$

$$[y] = [E][x] + [F][u] \tag{34}$$

where $A = A_1 D + A_2(1 - D)$

$$B = B_1 D + B_2(1 - D)$$

$$E = E_1 D + E_2(1 - D)$$

$$F = F_1 D + F_2(1 - D) \tag{35}$$

A_1, B_1, E_1, and F_1 represents ON-time state-space matrices and A_2, B_2, E_2, and F_2 represents OFF-time state-space matrices of converter with D as the duty cycle. From Eq. (2.27) the steady-state solution of the converter can be calculated.

$$\text{Under steady state } [\hat{x}] = 0,\ [x] = X_{ss}\ \&\ [u] = V_g \tag{36}$$

$$[A]X_{ss} + [B]V_g = 0 \tag{37}$$

$$[A]X_{ss} = -[B]V_g \tag{38}$$

$$[A]^{-1}[A]X_{ss} = -[A]^{-1}[B]V_g \tag{39}$$

$$X_{ss} = -[A]^{-1}[B]V_g \tag{40}$$

$$\text{Steady-state solution } X_{ss} = -A^{-1}BV_g \tag{41}$$

The proposed converter works in two modes based on power flow direction such as charger and discharger. In both the cases, the configuration will be different and needs to be modeled separately [2].

2.3.2 State-Space Model of Charger

The proposed converter is consisting of DC grid on source side and battery on the load side while operating as charger. So, the battery has been replaced by an equivalent resistor and a shunt capacitor for the ease of state-space analysis.

State-space model for Mode-1

In this mode, only switch S_1 is turned-on, S_2 and D_3 are turned-off. Then the equivalent circuit for mode-1 and corresponding state variables are as shown in Fig. 12

From Fig. 12

$$i_{C2} = i_{L1} \tag{42}$$

$$\frac{dv_{C2}}{dt} = \frac{1}{C_2} i_{L1} \tag{43}$$

$$i_{C1} = i_{L2} \tag{44}$$

$$\frac{dv_{C1}}{dt} = \frac{1}{C_1} i_{L2} \tag{45}$$

Applying KCL at node-2:

$$i_{L1} = i_{L2} + i_{C3} + I_0 \tag{46}$$

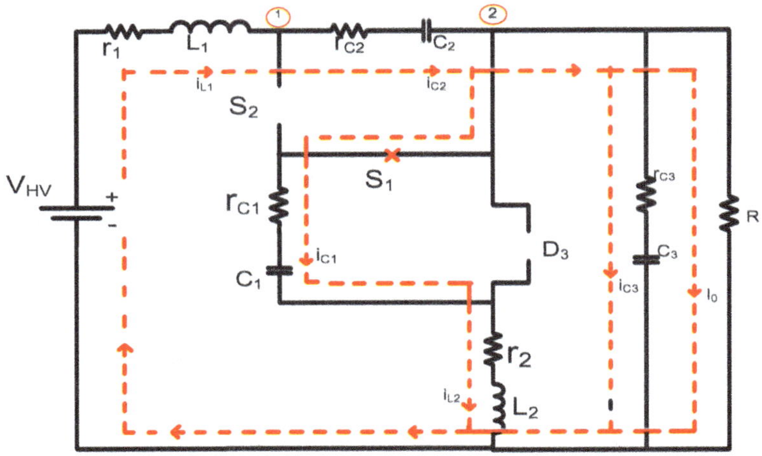

Fig. 12 Equivalent Circuit of charger during mode-1

$$\frac{dv_{C3}}{dt} = \frac{1}{C_3}\left(\frac{R}{R + r_{C3}}\right)i_{L1} - \frac{1}{C_3}\left(\frac{R}{R + r_{C3}}\right)i_{L2} - \frac{1}{C_3}\left(\frac{1}{R + r_{C3}}\right)v_{C3} \tag{47}$$

From Fig. 12, KVL through the path $\left(V_{HV} = V_g\right)$, L_1 , C_2, C_3

$$\frac{di_{L1}}{dt} = \frac{1}{L_1}V_g - \frac{1}{L_1}\left(r_1 + r_{C2} + \frac{Rr_{C3}}{R + r_{C3}}\right)i_{L1} + \frac{1}{L_1}\left(\frac{Rr_{C3}}{R + r_{C3}}\right)i_{L2} - \frac{1}{L_1}v_{C2} - \frac{1}{L_1}\left(\frac{R}{R + r_{C3}}\right)v_{C3} \tag{48}$$

From Fig. 12, KVL through the path S_1, C_1, L_2, C_3

$$L_2\frac{di_{L2}}{dt} = -(r_{C1} + r_2)i_{L2} - v_{C1} + v_{C3} + r_{C3}i_{C3} \tag{49}$$

$$\begin{aligned}\frac{di_{L2}}{dt} =& \frac{1}{L_2}\left(\frac{Rr_{C3}}{R + r_{C3}}\right)i_{L1} - \frac{1}{L_2}\left(r_{C1} + r_2 + \frac{Rr_{C3}}{R + r_{C3}}\right)i_{L2} - \frac{1}{L_2}v_{C1}\\ &+ \frac{1}{L_2}\left(\frac{R}{R + r_{C3}}\right)v_{C3}\end{aligned} \tag{50}$$

MODE-1:

$$\begin{bmatrix} \frac{di_{L1}}{dt} \\ \frac{di_{L2}}{dt} \\ \frac{dv_{C1}}{dt} \\ \frac{dv_{C2}}{dt} \\ \frac{dv_{C3}}{dt} \end{bmatrix} = \begin{bmatrix} -\frac{1}{L_1}\left(r_1 + r_{C2} + \frac{Rr_{C3}}{R+r_{C3}}\right) & \frac{1}{L_1}\left(\frac{Rr_{C3}}{R+r_{C3}}\right) & 0 & -\frac{1}{L_1} & -\frac{1}{L_1}\left(\frac{R}{R+r_{C3}}\right) \\ \frac{1}{L_2}\left(\frac{Rr_{C3}}{R+r_{C3}}\right) & -\frac{1}{L_2}\left(r_{C1} + r_2 + \frac{Rr_{C3}}{R+r_{C3}}\right) & -\frac{1}{L_2} & 0 & \frac{1}{L_2}\left(\frac{R}{R+r_{C3}}\right) \\ 0 & \frac{1}{C_1} & 0 & 0 & 0 \\ \frac{1}{C_2} & 0 & 0 & 0 & 0 \\ \frac{1}{C_3}\left(\frac{R}{R+r_{C3}}\right) & -\frac{1}{C_3}\left(\frac{R}{R+r_{C3}}\right) & 0 & 0 & -\frac{1}{C_3}\left(\frac{1}{R+r_{C3}}\right) \end{bmatrix} \begin{bmatrix} i_{L_1} \\ i_{L_2} \\ v_{C_1} \\ v_{C_2} \\ v_{C_3} \end{bmatrix}$$

$$+\begin{bmatrix} \frac{1}{L_1} \\ 0 \\ 0 \\ 0 \\ 0 \end{bmatrix} u$$

Output Voltage Matrix:

$$v = \left[\frac{Rr_{C3}}{R+r_{C3}} \; -\frac{Rr_{C3}}{R+r_{C3}} \; 0 \; 0 \; \frac{R}{R+r_{C3}} \right] \begin{bmatrix} i_{L1} \\ i_{L2} \\ v_{C1} \\ v_{C2} \\ v_{C3} \end{bmatrix} + [0]u$$

State-space model for Mode-2

In this mode, switch S_2 and diode D_3 will be conducting. Then the equivalent circuit for mode-2 will be as shown in Fig. 13.

From Fig. 13, applying KVL

$$V_g = r_1 i_{L1} + L_1 \frac{di_{L_1}}{dt} + r_{C1} i_{C1} + v_{C1} + r_2 i_{L2} + L_2 \frac{di_{L_2}}{dt} \tag{51}$$

$$V_g = r_1 i_{L1} + L_1 \frac{di_{L_1}}{dt} + r_{C2} i_{C2} + v_{C2} + r_2 i_{L2} + L_2 \frac{di_{L_2}}{dt} \tag{52}$$

$$V_g = r_1 i_{L1} + L_1 \frac{di_{L_1}}{dt} + r_{C2} i_{C2} + v_{C2} + r_{C3} i_{C3} + v_{C3} \tag{53}$$

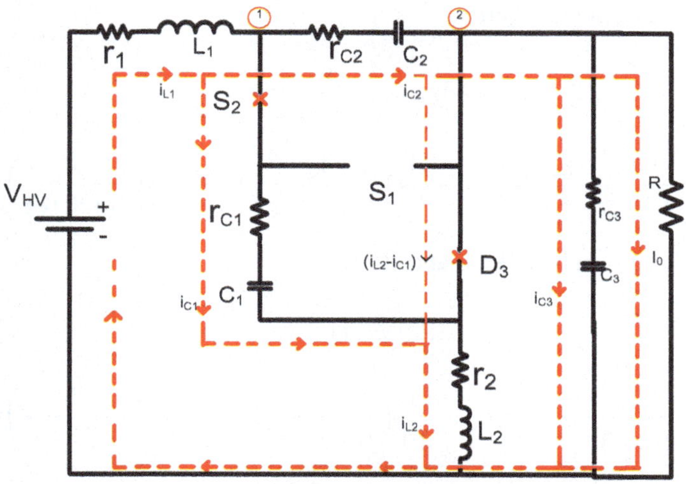

Fig. 13 Equivalent circuit of charger during mode-2

Applying KVL through the path S_2, C_1, D_3, C_2

$$r_{C2}i_{C2} + v_{C2} - v_{C1} - r_{C1}i_{C1} = 0 \tag{54}$$

Applying KCL at node-1:

$$i_{C2} + i_{C1} = i_{L1} \tag{55}$$

From Eqs. (52) and (53)

$$L_2\frac{di_{L2}}{dt} = r_{C3}i_{C3} - r_2 i_{L2} + v_{C3} \tag{56}$$

Applying KCL at node-2:

$$i_{C2} = i_{L2} - i_{C1} + i_{C3} + I_0 \tag{57}$$

$$\frac{dv_{C3}}{dt} = \frac{1}{C_3}\left(\frac{R}{R + r_{C3}}\right)i_{L1} - \frac{1}{C_3}\left(\frac{R}{R + r_{C3}}\right)i_{L2} - \frac{1}{C_3}\left(\frac{1}{R + r_{C3}}\right)v_{C3} \tag{58}$$

From Eq. (54)

$$r_{C2}i_{C2} - r_{C1}i_{C1} = v_{C1} - v_{C2} \tag{59}$$

From Eq. (55)

$$i_{C2} + i_{C1} = i_{L1} \tag{60}$$

Solving Eqs. (59) and (60)

$$i_{C2} = \frac{r_{C1}}{r_{C1} + r_{C2}}i_{L1} + \frac{1}{r_{C1} + r_{C2}}v_{C1} - \frac{1}{r_{C1} + r_{C2}}v_{C2} \tag{61}$$

$$\frac{dv_{C2}}{dt} = \frac{1}{C_2}\left(\frac{r_{C1}}{r_{C1} + r_{C2}}\right)i_{L1} + \frac{1}{C_2}\left(\frac{1}{r_{C1} + r_{C2}}\right)v_{C1} - \frac{1}{C_2}\left(\frac{1}{r_{C1} + r_{C2}}\right)v_{C2} \tag{62}$$

From Fig. 13 applying KCL at node-1:

$$i_{C1} = i_{L1} - i_{C2} \tag{63}$$

$$\frac{dv_{C1}}{dt} = \frac{1}{C_1}\left(\frac{r_{C2}}{r_{C1} + r_{C2}}\right)i_{L1} - \frac{1}{C_1}\left(\frac{1}{r_{C1} + r_{C2}}\right)v_{C1} + \frac{1}{C_1}\left(\frac{1}{r_{C1} + r_{C2}}\right)v_{C2} \tag{64}$$

From Eq. (53)

$$L_1\frac{di_{L1}}{dt} = V_g - r_1 i_{L1} - v_{C2} - v_{C3} - r_{C2}i_{C2} - r_{C3}i_{C3} \tag{65}$$

$$\frac{di_{L1}}{dt} = \frac{V_g}{L_1} - \frac{1}{L_1}\left(r_1 + \frac{r_{C1}r_{C2}}{r_{C1}+r_{C2}} + \frac{Rr_{C3}}{R+r_{C3}}\right)i_{L1} + \frac{1}{L_1}\left(\frac{Rr_{C3}}{R+r_{C3}}\right)i_{L2} - \frac{1}{L_1}\left(\frac{r_{C2}}{r_{C1}+r_{C2}}\right)v_{C1} - \frac{1}{L_1}\left(\frac{r_{C1}}{r_{C1}+r_{C2}}\right)v_{C2} - \frac{1}{L_1}\left(\frac{R}{R+r_{C3}}\right)v_{C3} \tag{66}$$

From Eq. (51)

$$L_2\frac{di_{L2}}{dt} = V_g - r_1 i_{L1} - L_1\frac{di_{L_1}}{dt} - r_{C1}i_{C1} - v_{C1} - r_2 i_{L2} \tag{67}$$

$$\frac{di_{L2}}{dt} = \frac{1}{L_2}\left(\frac{Rr_{C3}}{R+r_{C3}}\right)i_{L1} - \frac{1}{L_2}\left(r_2 + \frac{Rr_{C3}}{R+r_{C3}}\right)i_{L2} + \frac{1}{L_2}\left(\frac{R}{R+r_{C3}}\right)v_{C3} \tag{68}$$

MODE-2:

$$\begin{bmatrix} \frac{di_{L1}}{dt} \\ \frac{di_{L2}}{dt} \\ \frac{dv_{C1}}{dt} \\ \frac{dv_{C2}}{dt} \\ \frac{dv_{C3}}{dt} \end{bmatrix} =$$

$$\begin{bmatrix} -\frac{1}{L_1}\left(r_1 + \frac{r_{C1}r_{C2}}{r_{C1}+r_{C2}} + \frac{Rr_{C3}}{R+r_{C3}}\right) & \frac{1}{L_1}\left(\frac{Rr_{C3}}{R+r_{C3}}\right) & -\frac{1}{L_1}\left(\frac{r_{C2}}{r_{C1}+r_{C2}}\right) & -\frac{1}{L_1}\left(\frac{r_{C1}}{r_{C1}+r_{C2}}\right) & -\frac{1}{L_1}\left(\frac{R}{R+r_{C3}}\right) \\ \frac{1}{L_2}\left(\frac{Rr_{C3}}{R+r_{C3}}\right) & -\frac{1}{L_2}\left(r_2 + \frac{Rr_{C3}}{R+r_{C3}}\right) & 0 & 0 & \frac{1}{L_2}\left(\frac{R}{R+r_{C3}}\right) \\ \frac{1}{C_1}\left(\frac{r_{C2}}{r_{C1}+r_{C2}}\right) & 0 & -\frac{1}{C_1}\left(\frac{1}{r_{C1}+r_{C2}}\right) & \frac{1}{C_1}\left(\frac{1}{r_{C1}+r_{C2}}\right) & 0 \\ \frac{1}{C_2}\left(\frac{r_{C1}}{r_{C1}+r_{C2}}\right) & 0 & \frac{1}{C_2}\left(\frac{1}{r_{C1}+r_{C2}}\right) & -\frac{1}{C_2}\left(\frac{1}{r_{C1}+r_{C2}}\right) & 0 \\ \frac{1}{C_3}\left(\frac{R}{R+r_{C3}}\right) & -\frac{1}{C_3}\left(\frac{R}{R+r_{C3}}\right) & 0 & 0 & -\frac{1}{C_3}\left(\frac{1}{R+r_{C3}}\right) \end{bmatrix}$$

$$\begin{bmatrix} i_{L1} \\ i_{L2} \\ v_{C1} \\ v_{C2} \\ v_{C3} \end{bmatrix} + \begin{bmatrix} \frac{1}{L_1} \\ 0 \\ 0 \\ 0 \\ 0 \end{bmatrix} u$$

Output Voltage Matrix:

$$v = \begin{bmatrix} \frac{Rr_{C3}}{R+r_{C3}} & -\frac{Rr_{C3}}{R+r_{C3}} & 0 & 0 & \frac{R}{R+r_{C3}} \end{bmatrix} \begin{bmatrix} i_{L1} \\ i_{L2} \\ v_{C1} \\ v_{C2} \\ v_{C3} \end{bmatrix} + [0]u$$

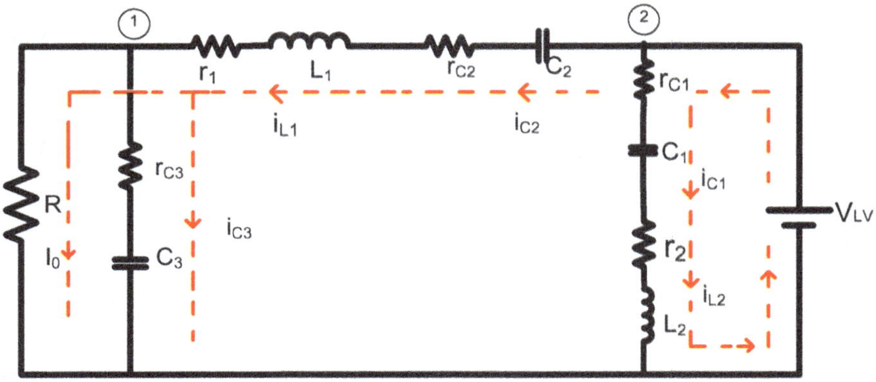

Fig. 14 Equivalent circuit of discharger during mode-1

2.3.3 State-Space Model for Discharger

The proposed converter is consisting of battery on source side and DC grid on the load side while operating as discharger. Therefore, DC grid has been replaced by an equivalent resistor and a shunt capacitor for the ease of state-space analysis

State-space model for Mode-1

In this mode, only diode D_1 is turned-on. Then the equivalent circuit for mode-1 will be as shown in Fig. 14

From Fig. 14

$$i_{C2} = i_{L1} \tag{69}$$

$$\frac{dv_{C2}}{dt} = \frac{1}{C_2} i_{L1} \tag{70}$$

$$i_{C1} = i_{L2} \tag{71}$$

$$\frac{dv_{C1}}{dt} = \frac{1}{C_1} i_{L2} \tag{72}$$

Applying KCL at node-1:

$$i_{L1} = i_{C3} + I_0 \tag{73}$$

$$i_{C3} = \frac{R}{R + r_{C3}} i_{L1} - \frac{1}{R + r_{C3}} v_{C3} \tag{74}$$

$$\frac{dv_{C3}}{dt} = \frac{R}{C_3(R + r_{C3})} i_{L1} - \frac{1}{C_3(R + r_{C3})} v_{C3} \tag{75}$$

Applying KVL in the right-hand side loop

$$\frac{di_{L2}}{dt} = \frac{V_{LV}}{L_2} - \frac{(r_{C1} + r_2)}{L_2} i_{L2} - \frac{1}{L_2} v_{C1} \tag{76}$$

Applying KVL through the path V_{LV},C_2,L_1 ,C_3

$$V_{LV} = r_{C2} i_{C2} + v_{C2} + r_1 i_{L1} + L_1 \frac{di_{L1}}{dt} + r_{C3} i_{C3} + v_{C3} \tag{77}$$

$$\frac{di_{L1}}{dt} = \frac{V_{LV}}{L_1} - \frac{1}{L_1}\left(r_1 + r_{C2} + \frac{R r_{C3}}{R + r_{C3}}\right) i_{L1} - \frac{1}{L_1} v_{C2} - \frac{R}{L_1(R + r_{C3})} v_{C3} \tag{78}$$

Output Voltage:

$$V_0 = r_{C3} i_{C3} + v_{C3}$$

$$V_0 = \frac{R r_{C3}}{R + r_{C3}} i_{L1} + \frac{R}{R + r_{C3}} v_{C3} \tag{79}$$

MODE-1:

$$\begin{bmatrix} \frac{di_{L1}}{dt} \\ \frac{di_{L2}}{dt} \\ \frac{dv_{C1}}{dt} \\ \frac{dv_{C2}}{dt} \\ \frac{dv_{C3}}{dt} \end{bmatrix} = \begin{bmatrix} \frac{-1}{L_1}\left[r_1 + r_{C2} + \frac{R r_{C3}}{R + r_{C3}}\right] & 0 & 0 & \frac{-1}{L_1} & \frac{-R}{L_1(R + r_{C3})} \\ 0 & \frac{-(r_{C1} + r_2)}{L_2} & \frac{-1}{L_2} & 0 & 0 \\ 0 & \frac{1}{C_1} & 0 & 0 & 0 \\ \frac{1}{C_2} & 0 & 0 & 0 & 0 \\ \frac{R}{C_3(R + r_{C3})} & 0 & 0 & 0 & \frac{-1}{C_3(R + r_{C3})} \end{bmatrix} \begin{bmatrix} i_{L1} \\ i_{L2} \\ v_{C1} \\ v_{C2} \\ v_{C3} \end{bmatrix} + \begin{bmatrix} \frac{1}{L_1} \\ \frac{1}{L_2} \\ 0 \\ 0 \\ 0 \end{bmatrix} u$$

Output voltage Matrix:

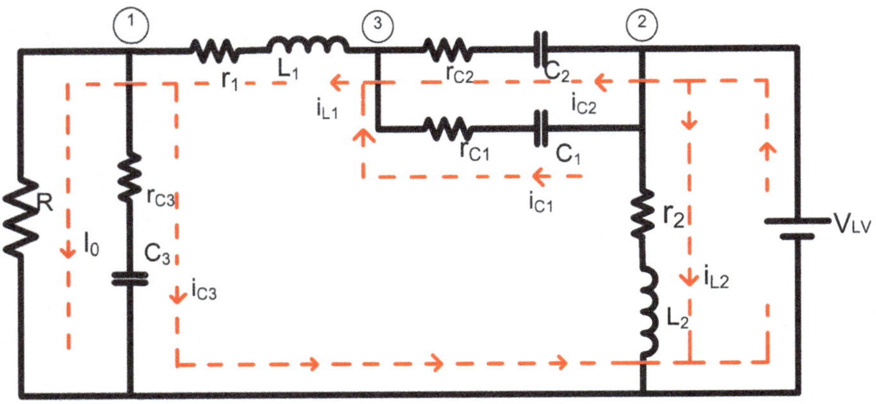

Fig. 15 Equivalent circuit of discharger during mode-2

$$V_0 = \begin{bmatrix} \frac{Rr_{C3}}{R + r_{C3}} & 0 & 0 & 0 & \frac{R}{R + r_{C3}} \end{bmatrix} \begin{bmatrix} i_{L1} \\ i_{L2} \\ v_{C1} \\ v_{C2} \\ v_{C3} \end{bmatrix} + [0]u$$

State-space model for Mode-2

In this mode switch S_3 and diode D_2 will be conducting. Then the equivalent circuit for mode-2 will be as shown in Fig. 15.

Applying KCL at node-1:

$$i_{L1} = i_{C3} + I_0 \tag{80}$$

$$\frac{dv_{C3}}{dt} = \frac{R}{C_3(R + r_{C3})} i_{L1} - \frac{1}{C_3(R + r_{C3})} v_{C3} \tag{81}$$

Applying KVL in the right-hand side loop

$$r_2 i_{L2} + v_{L2} = V_{LV} \tag{82}$$

$$\frac{di_{L2}}{dt} = \frac{-r_2}{L_2} i_{L2} + \frac{V_{LV}}{L_2} \tag{83}$$

From Fig. 15 applying KVL through the path r_{C2}, C_2, C_1, r_{C1}

$$r_{C2} i_{C2} + v_{C2} = r_{C1} i_{C1} + v_{C1} \tag{84}$$

Applying KCL at node-3:

$$i_{C2} + i_{C1} = i_{L1} \tag{85}$$

Solving the above two Eqs. (84) and (85)

$$\frac{dv_{C2}}{dt} = \frac{r_{C1}}{C_2(r_{C1} + r_{C2})} i_{L1} + \frac{1}{C_2(r_{C1} + r_{C2})} v_{C1} - \frac{1}{C_2(r_{C1} + r_{C2})} v_{C2} \tag{86}$$

From Eq. (85)

$$i_{C1} = i_{L1} - i_{C2} \tag{87}$$

$$\frac{dv_{C1}}{dt} = \frac{r_{C2}}{C_1(r_{C1} + r_{C2})} i_{L1} - \frac{1}{C_1(r_{C1} + r_{C2})} v_{C1} + \frac{1}{C_1(r_{C1} + r_{C2})} v_{C2} \tag{88}$$

Applying KVL through the path V_{LV},C_2,L_1 ,C_3

$$V_{LV} = r_{C2} i_{C2} + v_{C2} + r_1 i_{L1} + L_1 \frac{di_{L1}}{dt} + r_{C3} i_{C3} + v_{C3} \tag{89}$$

$$\frac{di_{L1}}{dt} = \frac{V_{LV}}{L_1} - \frac{1}{L_1}\left(r_1 + \frac{r_{C1} r_{C2}}{r_{C1} + r_{C2}} + \frac{R r_{C3}}{R + r_{C3}}\right) i_{L1} - \frac{r_{C2}}{L_1(r_{C1} + r_{C2})} v_{C1} - \frac{r_{C1}}{L_1(r_{C1} + r_{C2})} v_{C2} - \frac{R}{L_1(R + r_{C3})} v_{C3} \tag{90}$$

Output Voltage:

$$V_0 = r_{C3} i_{C3} + v_{C3} \tag{91}$$

$$V_0 = \frac{R r_{C3}}{R + r_{C3}} i_{L1} + \frac{R}{R + r_{C3}} v_{C3} \tag{92}$$

MODE-2:

$$\begin{bmatrix} \frac{di_{L1}}{dt} \\ \frac{di_{L2}}{dt} \\ \frac{dv_{C1}}{dt} \\ \frac{dv_{C2}}{dt} \\ \frac{dv_{C3}}{dt} \end{bmatrix} = \begin{bmatrix} \frac{-1}{L_1}\left[r_1 + \frac{r_{C1} r_{C2}}{r_{C1} + r_{C2}} + \frac{R r_{C3}}{R + r_{C3}}\right] & 0 & \frac{-r_{C2}}{L_1(r_{C1} + r_{C2})} & \frac{-r_{C1}}{L_1(r_{C1} + r_{C2})} & \frac{-R}{L_1(R + r_{C3})} \\ 0 & \frac{-r_2}{L_2} & 0 & 0 & 0 \\ \frac{r_{C2}}{C_1(r_{C1} + r_{C2})} & 0 & \frac{-1}{C_1(r_{C1} + r_{C2})} & \frac{1}{C_1(r_{C1} + r_{C2})} & 0 \\ \frac{r_{C1}}{C_2(r_{C1} + r_{C2})} & 0 & \frac{1}{C_2(r_{C1} + r_{C2})} & \frac{-1}{C_2(r_{C1} + r_{C2})} & 0 \\ \frac{R}{C_3(R + r_{C3})} & 0 & 0 & 0 & \frac{-1}{C_3(R + r_{C3})} \end{bmatrix}$$

$$\begin{bmatrix} i_{L1} \\ i_{L2} \\ v_{C1} \\ v_{C2} \\ v_{C3} \end{bmatrix} + \begin{bmatrix} \frac{1}{L_1} \\ \frac{1}{L_2} \\ 0 \\ 0 \\ 0 \end{bmatrix} u$$

Output voltage Matrix:

$$V_0 = \begin{bmatrix} \frac{Rr_{C3}}{R + r_{C3}} & 0 & 0 & 0 & \frac{R}{R + r_{C3}} \end{bmatrix} \begin{bmatrix} i_{L1} \\ i_{L2} \\ v_{C1} \\ v_{C2} \\ v_{C3} \end{bmatrix} + [0]u$$

3 Design of Converter

The proposed converter is intended for use in telecommunication applications in which 48 V DC supply is the requirement. The DC grid can be designed for this voltage to suit this purpose and hence this value is the converter DC grid voltage. Most of the battery manufacturers will produce batteries with 12, 24 V, etc. So, the battery voltage is selected to be 12 V. The laboratory prototype is made for 100 W and this can be redesigned for higher power requirements by following the same approach.

In the following section, the converter components are calculated by considering charger operation as shown in Fig. 16 However, the design will not change for the same analysis of discharger operation.

For designing the converter components, the following assumptions have been made.

1. The switches and the diodes in the circuit are ideal.
2. The passive components of the circuit are linear, time-invariant and also independent of frequency.
3. The proposed converter operates in continuous current mode (CCM).
4. The component non-idealities are zero.

Based on the above assumptions, the time-domain analysis has been carried out to design the converter components.

Using power balance

$$V_{HV} I_{L1} = V_{LV} I_0 \tag{93}$$

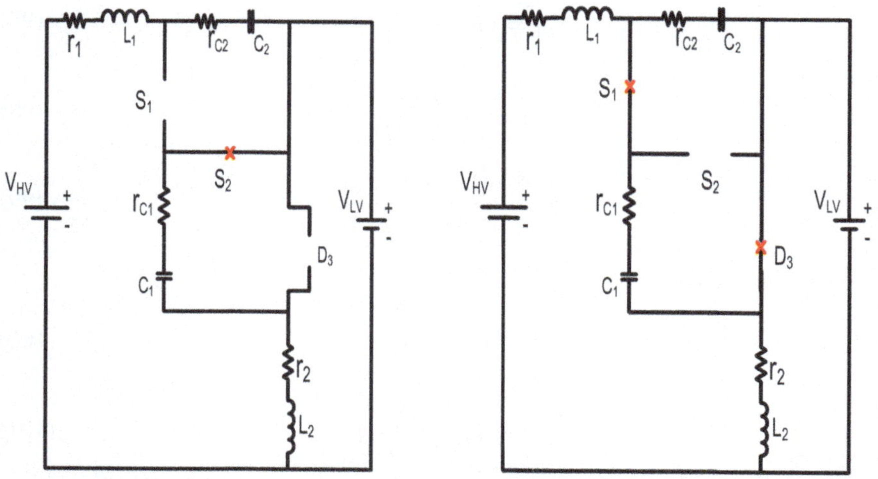

(a) Equivalent Circuit for Mode-1 (b) Equivalent Circuit for Mode-2

Fig. 16 Equivalent circuits of charger in CCM mode

$$V_{HV}I_{L1} = V_{HV}\left(\frac{D}{1+D}\right)I_0 \tag{94}$$

$$I_{L1} = \left(\frac{D}{1+D}\right)I_0 \tag{95}$$

For finding inductor expressions apply volt-sec balance [2] for a switching cycle across desired inductance. From Fig. 17, it is shown D is duty cycle and T_s is switching time-period.

From the Fig. 16b

$$L_2 = V_{LV}\frac{(1-D)}{f_s \Delta i_{L2}} \tag{96}$$

From Fig. 16a

Fig. 17 Current ripple across inductor

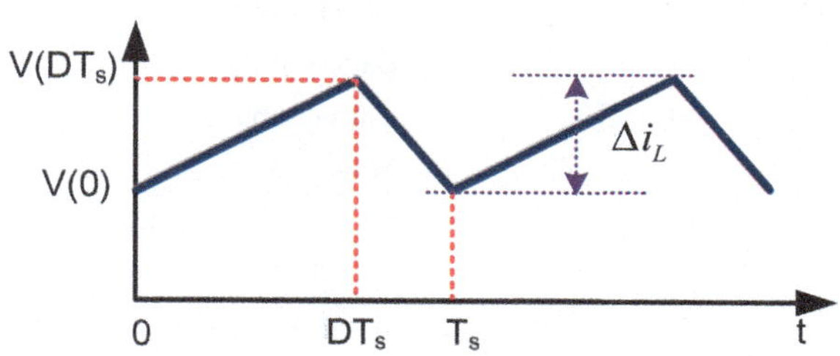

$$v_{L1} = v_{HV} - v_{C2} - v_{LV} \tag{97}$$

$$v_{L1} = v_{LV}\left(\frac{1 - D^2}{D}\right) \tag{98}$$

$$L_1 = \frac{(1 - D)}{D}\frac{v_{LV}}{f_s di_{L1}} \tag{99}$$

From Fig. 16a

$$i_{C2} = i_{L1} \tag{100}$$

$$C_1 = C_2 = I_0\left(\frac{D}{1 + D}\right)\frac{D}{f_s \Delta V_{C2}} \tag{101}$$

Applying KCL for Fig. 16a

$$i_{C3} = i_{L1} - i_{C1} - I_0 \tag{102}$$

$$i_{C3} = i_{L1}\left(\frac{2 + D}{2D}\right) \tag{103}$$

$$C_3 = \frac{V_0}{R}\left(\frac{2D + D^2}{2 + 2D}\right)\frac{1}{f_s \Delta V_{C3}} \tag{104}$$

Neglecting higher order terms,

$$C_3 = \left(\frac{2D}{2 + 2D}\right)\frac{V_0}{R f_s \Delta V_{C3}} \tag{105}$$

Using the specifications of the converter as presented in Table 2, the component values have been calculated as shown in Table 3

Table 2 Converter Specifications

Power rating	100 W
DC grid voltage	48 V
Battery voltage	12 V
Current ripple	≤10%
Voltage ripple	≤5%
Switching frequency	50 kHz

Table 3 Calculation of converter components

$L_1 = \frac{(1-D)}{D}\frac{V_{LV}}{f_s di_{L1}}$	50 μH
$L_2 = V_{LV}\frac{(1-D)}{f_s \Delta i_{L2}}$	63 μH
$C_1 = I_0\left(\frac{D}{1+D}\right)\frac{D}{f_s \Delta V_{C1}}$	47 μF
$C_2 = I_0\left(\frac{D}{1+D}\right)\frac{D}{f_s \Delta V_{C2}}$	47 μF
$C_3 = \left(\frac{2D}{2+2D}\right)\frac{V_0}{R f_s \Delta V_{C3}}$	180 μF

3.1 Steady-State Performance of Converter

The fifth-order bidirectional DC-DC converter has been simulated with the designed component values of Table 3 using PSIM simulator with an objective of knowing the system behavior under nominal conditions. The steady-state performance in charger operation is as shown in Fig. 18 and in discharger operation is shown in Fig. 19

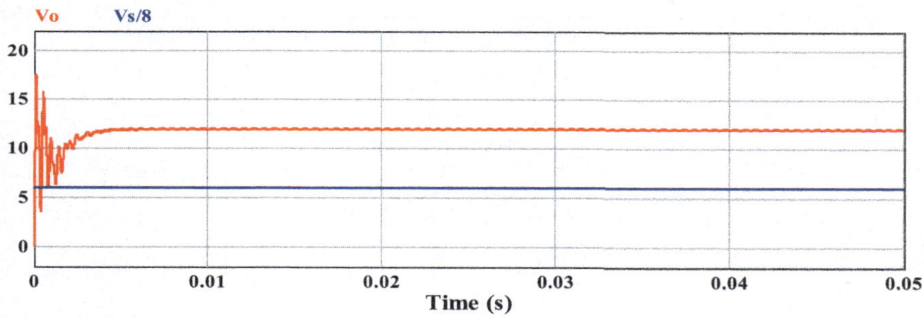

Fig. 18 Charger steady-state performance

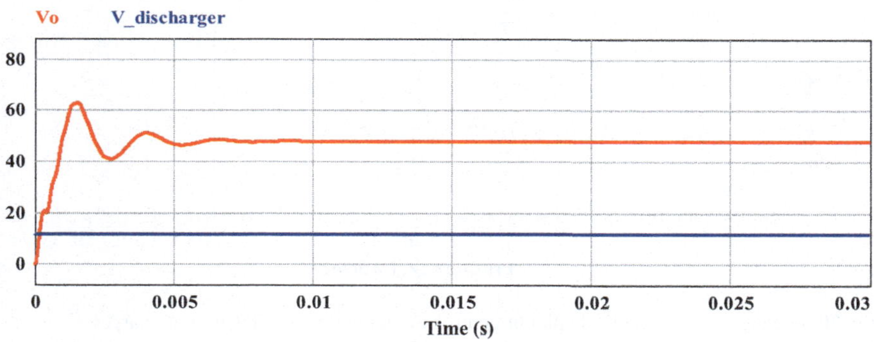

Fig. 19 Discharger steady-state performance

Charger Steady-state Performance

See Fig. 18.

Discharger Steady-state Performance

See Fig. 19.

4 Compensator Design for Charger

Using the state-space and discrete-time modeling of the converter as discussed in previous chapter, we have obtained control-to-output transfer function of the charger as given in Eq. (4.23). The frequency response of this transfer function will help to identify system behavior [12]. The Bode plot for control-to-output transfer function is obtained as shown in Fig. 20.

Control-to-output transfer function for charger operation (Fig. 21)

$$G_{vd\_c}(z) = \frac{0.7975z^4 - 1.783z^3 + 1.489z^2 - 0.5036z + 0.05688}{z^5 - 3.718z^4 + 5.495z^3 - 3.971z^2 + 1.358z - 0.1603} \quad (106)$$

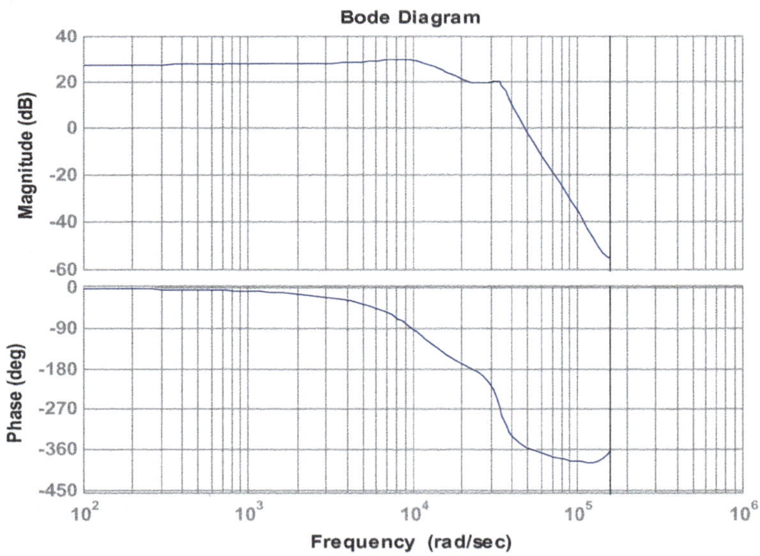

Fig. 20 Frequency response Bode plot of control-to-output transfer function of charger

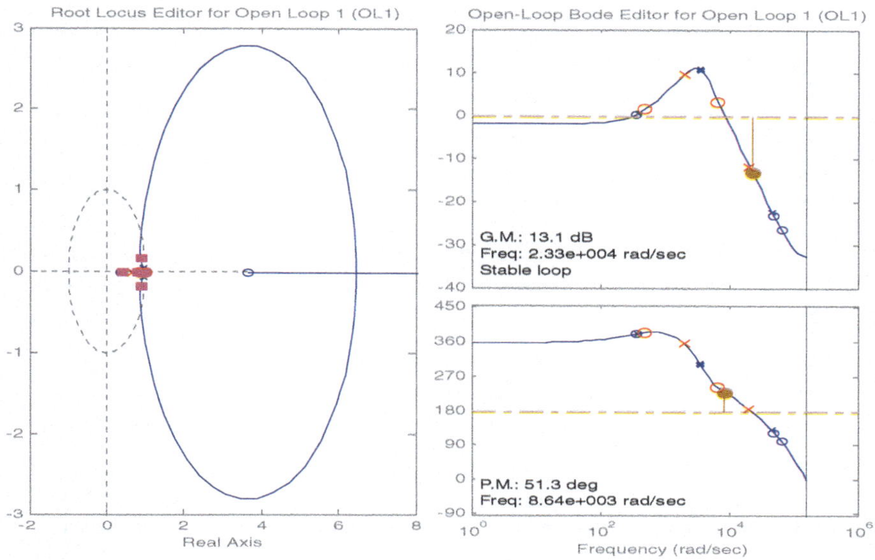

Fig. 21 Designing digital compensator to control-to-output transfer function of charger in SISO tool

5 Digital Compensator Design of Discharger

Compensator for discharger is designed in the same way as charger operation. Using the state-space and discrete-time modeling of the converter as discussed in previous chapter, we have obtained control-to-output transfer function of the discharger as given in Eq. (5.1). The frequency response of this transfer function will help to identify system behaviors [2, 12]. The Bode plot for discharger control-to-output transfer function is obtained as shown in Fig. 22.

Control-to-output transfer function for discharger operation (Fig. 23)

$$G_{vd\_d}(z) = \frac{0.05236z^4 + 0.1731z^3 + 0.1116z^2 + 0.01478z - 0.0002303}{z^5 - 3.368z^4 + 4.706z^3 - 3.31z^2 + 1.087z - 0.1011} \quad (107)$$

6 Conclusion

The designed converter is simulated on PSIM Simulator. The various current and voltage waveforms have been compared with the bidirectional DC-DC converter without complimentary switching, which shows that the designed converter operates in the soft switching mode for increasing the efficiency with reduction in switching losses. Linear or small-signal models of the converter are required to

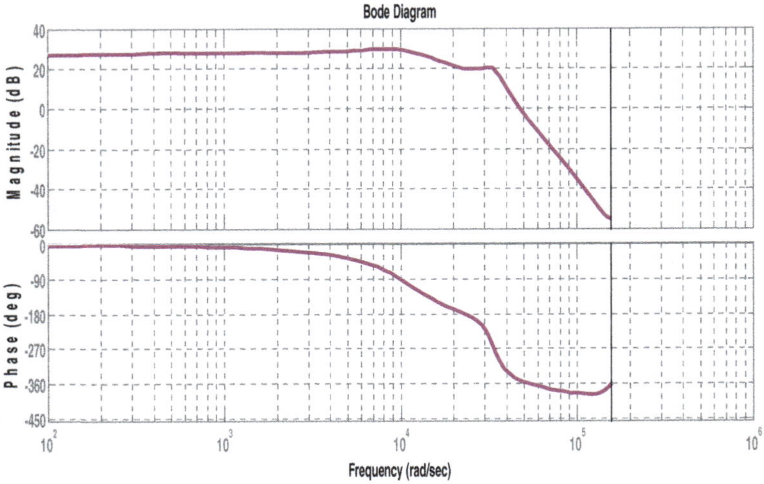

Fig. 22 Frequency response Bode plot of control-to-output transfer function of Discharger

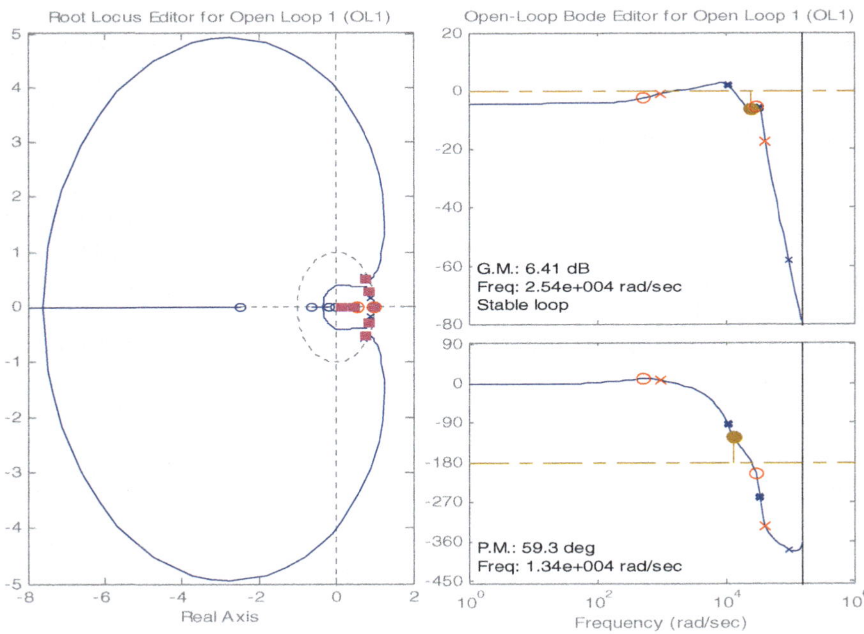

Fig. 23 Designing digital compensator of control-to-output transfer function of discharger in SISO tool

design the digital voltage-mode controller. For designing the controller, the standard pole-placement technique (SPPT) is used here. Here for designing the controller MATLAB-SISOTOOL interactive platform is used, where controller's poles and zero locations can be easily adjusted under stable conditions such as gain margin >6 *dB*, and phase margin: $45^{o} \sim 75^{o}$.

References

1. M. Veerachay, V.K. Tewari, N. Anandh, Two input boosting Dc-Dc converter for Dc grid application, in *Proceeding of the IEEE Conference on Green Computing Communication and Conservation of Energy*, vol. 54 (2013), pp. 574–579
2. N. Saisrinivas, Y. Nikhil, S.M. Shyni, Bi-Directional Ev Charger For Grid Connected Electric Vehicle Applications. Int. J. Pure Appl. Math. **120**(6), 983–994 (2018)
3. J. Dou, X. Bizhong, W. Zheng, A novel bidirectional onboard charger for electric vehicle, in *Advances in Mechanical and Electronic Engineering* (Springer, Berlin, Heidelberg, 2012), pp. 521–527
4. A. Rahman, M. Ali, M. Aamir, M. Mumtaz, H.S. Khan, H.S., A. Waqar, A design and analysis of efficient battery charging/discharging for V2G and G2V application, in *Proceeding of the 2018 2nd International Conference on Energy Conservation and Efficiency (ICECE)* (2018), pp. 6–13
5. K.W. Hu, P.H. Yi, C.M. Liaw, An EV SRM Drive Powered By Battery/Supercapacitor With G2v And V2h/V2g Capabilities. IEEE Trans. Industr. Electron. **62**(8), 4714–4727 (2015)
6. S. Jalbrzykowski, T. Citko, A bidirectional DC-DC converter for renewable energy systems. Bull Polish Acad Sci Techn Sci **57**(4) (2009)
7. J. Dou, B. Xia, W. Zheng, A novel bidirectional onboard charger for electric vehicle, in *Advances in Mechanical and Electronic Engineering* (Springer, Berlin, Heidelberg, 2012), pp. 521–527
8. M.M. Breve, V. Leite, Control of a bidirectional single-phase grid interface for electric vehicles, in *Ibero-American Congress on Information Management and Big Data* (Springer, Cham, 2019), pp. 285–299
9. V. Monteiro, J. Catalao, T. Sousa, J. Pinto, M. Mezaroba, J. Afonso, Improved voltage control for the electric vehicle operation in V2H mode as an off-line UPS in the context of smart homes. EAI Endorsed Trans Energy Web **7**(25) (2020)
10. A. Soares, A. Gomes, C.H. Antunes, C. Oliveira, A customized evolutionary algorithm for multi objective management of residential energy resources. IEEE Trans. Ind. In format **13**(2), 492–501 (2017)
11. S.K. Panda, A. Ghosh, A computational analysis of interfacing converters with advanced control methodologies for microgrid application. Technol. Econ. Smart Grids Sustain. Energy **5**(1), 1–18 (2020)
12. A. Toloei, Z. Mona, G. Reza, M. Fazel, A comparative analysis of star identification algorithms. Astrophys. Space Sci. (2020)

Photovoltaic Inverter Model in Simulink

Mostafa Al-Gabalawy

1 Introduction

The 2020 National Renewable Energy Lab (NREL) Solar District Cup is a collegiate design competition that challenges undergraduates to design a realistic and buildable solar photovoltaic (PV) plus battery storage system for one of several urban districts. Multiple Lane Department of Electrical Engineering and Computer Science Seniors participated in this competition in lieu of their capstone as competitors in the New Mexico State University (NMSU) district division, and placed second overall. NMSU wishes to reduce their energy impact and maximize utility savings by investing in a commercial solar PV plus battery storage system. To this end, the team designed two different solar PV installations on the NMSU campus that collectively supply approximately 30% of the campus' yearly energy consumption. The installation of interest to this project is a 278.2 kW-DC roof-mounted installation on the Corbett Center Student Union. This installation, like most PV plus storage systems, necessitates the use of a power inverter, of which there are two in the Corbett Center installation, as can be seen in the one-line diagram in the appendix. It is important to note that local zoning laws in the region of Las Cruces, New Mexico that the university is located presently prohibit net metering. Due to this situation, the overall campus PV system was designed such that power generation does not result in net power exports back into the grid. However, the PV plus storage system must be designed to be 'grid ready' and is interconnected onto the Geothermal Substation of the Western Interconnection region. Like in many jurisdictions, net-metering is slated to gain legislative ground in the coming years. If and when net metering is permitted, it will require that the utility that services the campus credits the campus for the solar power that is supplied to the grid. When it begins net metering, the Corbett Center Student Union roof-mounted installation's power inverter; being the electronic system that

M. Al-Gabalawy (✉)
Pyramids Higher Institute for Engineering and Technology, Giza, Egypt
e-mail: mostafagabalawy@gmail.com

V. K. Singh et al. (eds.), *Renewable Energy and Future Power Systems*, Energy Systems in Electrical Engineering, https://doi.org/10.1007/978-981-33-6753-1_6

converts the DC power generated by the PV array into AC usable by the grid, must output power that has been processed such that it is sufficiently cleared of distorting harmonic contents. Thus, the design of the inverter must seek to minimize both real power losses and total harmonic distortion (THD). This chapter presents a "reverse engineering" and redesign of one of the two SMA Sunny Tri-power Core1 50-US/62-US grid-tie inverters utilized in the PV installation designed for the Corbett Student Union rooftop by the team in the 2020 NREL Solar District Cup. This inverter is boxed in red in the one-line diagram in the appendix.

2 Circuit Topology and Devices

The chief design principle in question for the creation of any power electronic converter is the circuit topology. The team's PV inverter utilizes power MOSFETs as the primary active circuit components within the H-bridge, specifically the Mouser Silicon Carbide Power MOSFET C2M0025120D. Unlike current-controlled switching elements such as BJTs, MOSFETs are voltage-controlled, providing the advantage of considerably lower overall real power losses, and thereby lower costs associated with their operation as a power device. Additionally, MOSFETs with SiC substrate were selected due to the fact that they "switch faster, maintain a higher current density, and possess lower on-state impedance" compared to even most insulated-gate BJTs [1].

As shown in Fig. 1, before the H-bridge section of the inverter is a DC-link capacitor placed in parallel over the photovoltaic array to act as a "pre-inverter filter", in an attempt to smooth the inherently volatile and noisy DC signal of the PV array and limit its fluctuations even under heavy current absorption. This capacitor essentially acts as a filter to reduce current ripple [2]. Following the H-bridge, the output of the circuit, still being primarily a distorted sinusoid in nature, (i.e. possessing various harmonics) requires an extensive filtering scheme to produce a more pure sinusoidal output. This is accomplished primarily with a resonant filter consisting of an inductor

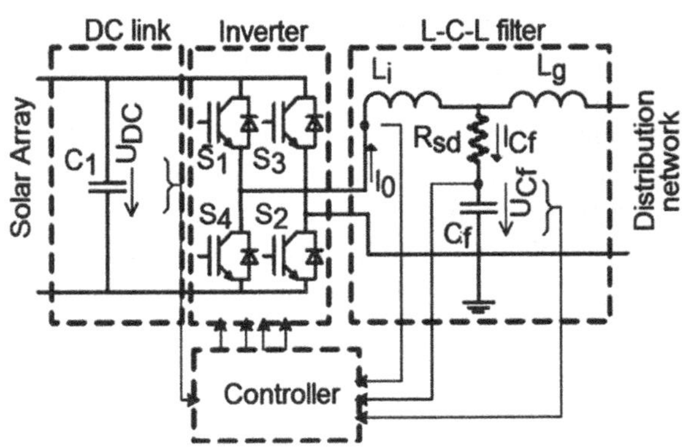

Fig. 1 General One-Line Diagram of a Grid-Tie Inverter (Single Phase) [2]

in series with the output terminal, and a capacitor in parallel with the output terminals. This is called an LC filter, which acts similar to the DC link, but instead functions on the output of the inverter.

Downsides of this LC filter include a marginal attenuation of the output current and the potential for resonances and unstable states in the system [3]. Following the LC filter, the voltage must be stepped up via a high frequency transformer to a little above the grid voltage before it can be connected. Because of the fact that Las Cruces, New Mexico does not currently allow net metering, the PV system and its corresponding inverter were modeled as operating in "Island Mode", and the modeling of the interconnection to the grid was neglected for the purpose of this analysis, although the inclusion of it in future works would be relatively simple.

An additional measure included in the inverter design, to work in tandem with the H-bridge, is a pulse width modulation (PWM) scheme. Naturally, the frequency of the output must be a practically pure 60 Hz, and therefore must also be synchronized with grid phase delay.

3 Device Protection and Calculations

The single inverter in the Corbett Hall PV System simulated by the team is fed by 12 strings of 16 PV modules. By referring to the specification sheet of the selected solar module, [4], the nominal, maximum, and worst case scenario specifications for the input of the solar array into the inverter were calculated utilizing the data for the CS32-420 PB-AG Module. By simulation, these values were confirmed to be successfully input into the inverter model via the PV array Simulink component.

A relatively unintuitive component of power electronics design is that of filter parameter selection. In order to filter the noisy sinusoidal PWM signal output by the h-bridge containing harmonics, a variety of protection filter schemes were utilized as described in Sect. 2. As described previously, a low pass RLC filter was implemented, utilizing a capacitor connected in parallel with the resistive load, and an inductor connected in series with the parallel combination. The overall thevenin equivalent of this is then connected to the H-bridge. The concept of quality factor, Q is described. This factor can be used in calculations for the various filter components of the inverter. From [5]:

$$Q = \frac{V_{Lp}}{V_{Lp(fund)}}$$

where, V_{Lp} is the peak voltage across the load, and $V_{Lp(fund)}$ is the peak of the fundamental voltage component across the load. From [6], a formulation for the inductance and capacitance values for the LC filter of a three-phase PV inverter can be given as:

$$L_f = \frac{0.1V^2}{2\pi f_0 P}$$

And

$$C_f = \frac{0.05P}{2\pi f_0 V^2}$$

A quality factor, of 1.07 was calculated using results from the simulation. Values of L_f and C_f were calculated to be 76 mH and 46 μF, respectively. Although no breakers were used throughout the simulation, in practice, breakers would be installed to provide layers of protection for devices that could be affected by a short circuit or arc flash.

Another possible mode of protection to prevent the MOSFETs from malfunctioning is to set a limit on the values of change in current and voltage per time, or $\frac{dv}{dt}$ and $\frac{di}{dt}$ limits. This can be accomplished via a snubber circuit, which can include multiple resistors and inductors and a capacitor. Complex snubber circuits utilize these passive components to introduce convergences within the turn on $\frac{di}{dt}$ value and turn off $\frac{dv}{dt}$ value. The RLC circuit is typically made to be critically damped to decrease effects of oscillations. With complex snubber circuits, resistors can also be placed across the main capacitor to avoid the capacitor from discharging across the MOSFET. In the team's simulation, the chosen universal bridge utilizes a simple RC snubber in parallel with each switching device.

3.1 Conduction Losses

According to [7], the average value of the conduction losses for a power MOSFET, denoted as $P_{C,M}$, can be calculated utilizing the following equation:

$$P_{C,M} = R_{DS,\text{on}} . I_{D,rms}^2$$

Thus, referencing the datasheet for the selected Mouser SiC Power MOSFET [3], the conduction losses for the 4 MOSFETs utilized in the H-bridge can be computed as shown below, ensuring to account for current division on input current into the H-bridge:

$$P_{\text{ave,conduction}} = \left(\frac{I_{\text{mar}}}{2}\right)^2 * R_{\text{ds,on}}$$

This value will later be used in the thermal characterization of the device, and motivate the selection of a heat-sink.

3.2 Switching Losses

In addition to conduction losses, switching losses are also prevalent in the circuit. The MOSFETs used incur losses due to the nonlinearities present during the mode switching. A reference document was used to aid in the calculations of switching losses for MOSFETs, [7]. Utilizing the datasheets from the solar modules and MOSFETs used, the information required to calculate energy lost during switching can be observed via graphs and tables. The first step in the process is to analyze the Gate Charge Characteristic graph (Fig. 2), which details the relationship between gate charge and gate-source voltage. Using the graph, the plateau voltage can be seen to hold a value of 10 V ($V_{Plateau}$).

Due to the Miller effect, the drive voltage must be taken at a higher value than that in the plateau region, taken as 13 V (V_{Dr}) which is used to turn on the MOSFET [7]. The next step is to take note of the gate resistance from the MOSFET datasheet (R_g= 1.1Ω) and then calculate the gate currents during rise and fall times:

$$I_{G,on} = \frac{V_{Dr} - V_{\text{plateau}}}{R_g}$$

$$I_{G,\text{off}} = \frac{-V_{\text{plateau}}}{R_g}$$

Following the gate current calculations, the time it takes for the voltage needs to be calculated. To start this calculation, the capacitance needs to be observed from the graphs that plot capacitance versus drain-source voltage (Fig. 3). Follow the graph

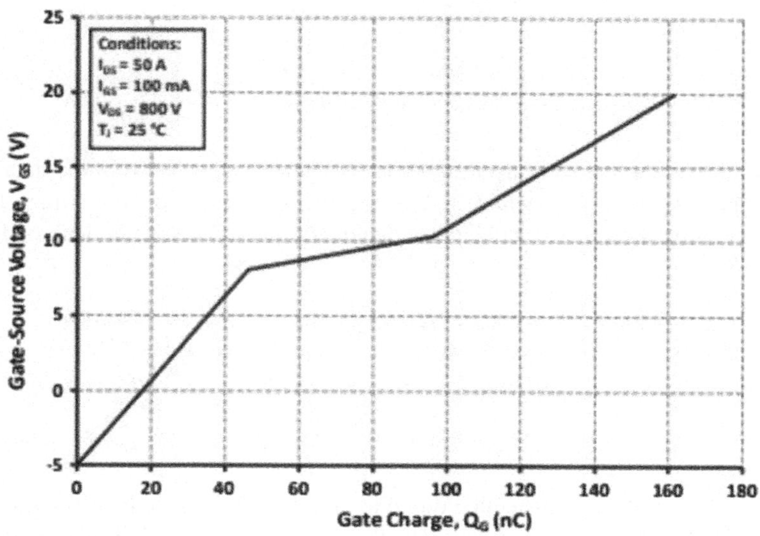

Fig. 2 Gate charge characteristic

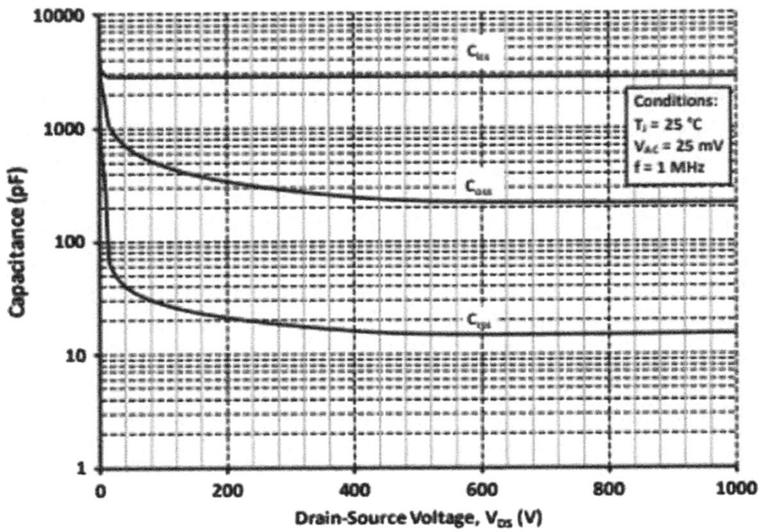

Fig. 3 Capacitances versus Drain-Source Voltage

given the supply voltage of 150 V and find that the capacitance value (*Crss*) is 25 pF. Half of the supply voltage produces capacitance (*Coss*) of around 550 pF.

After observing the capacitance values, the times it takes for the voltages to drop and rise along with corresponding averages are calculated:

$$t_{OFFv,1} = \left(V_{DD} - R_{DS,on} * I_{D,on}\right) * \frac{25pF}{I_{G,on}}$$
$$t_{OFFv,2} = \left(V_{DD} - R_{DS,on} * I_{D,on}\right) * \frac{550pF}{I_{G,on}}$$

$$t_{ONv,1} = \left(V_{DD} - R_{DS,on} * I_{D,on}\right) * \frac{25pF}{I_{G,off}}$$
$$t_{ONv,2} = \left(V_{DD} - R_{DS,on} * I_{D,on}\right) * \frac{550pF}{I_{G,\text{off}}}$$

Due to the fact that the current fall and rise times are included in the datasheet, they do not need to be re-calculated. After gathering the corresponding times, losses from switching need to be calculated. The equations used below are gathered from [7].

$$E_{loss,on} = V_{DD} * I_{D,on}.\frac{(i_{\text{risel time}}) + (v_{\text{fall time}})}{2} + Q_{rr}.V_{DD}$$

$$E_{\text{loss,off}} = V_{DD}.I_{D,\text{off}}.\frac{(i_{\text{fall time}}) + (v_{\text{risel time}})}{2}$$

Due to the built-in freewheeling diode in the MOSFET, some losses will need to be estimated and included into the overall switching losses within the circuit. The equation for total power losses due to switching is as follows:

$$\left(E_{loss,on} + E_{loss,off}\right).f_{sw} + \frac{1}{4} * Q_{rr}.V_{DD}.f_{sw}$$

3.3 Thermal Characterization

Generally, the goal of the heat-sink design is to limit the thermal impedance from the junction of the device to the ambient air. Therefore, the selection of a heat-sink onto which the device will be mounted is primarily dependent on the thermal impedance of the sink to the ambient air, $R_{\theta sa}$. In order to analyze the thermal characteristics of the overall device, a "thermal circuit" can be created with thermal impedances supplied by the average power of the system to find the thermal impedance between the case and the ambient air. After finding the case to ambient thermal impedance, this along with other various parameters can be utilized to find a suitable commercially available heat sink. By referring to the specification sheet of the selected Mouser C2M0025120D SiC Power MOSFET [3], the upper limit of the thermal impedance from the sink to the ambient air can be estimated with the equations and corresponding parameters as follows:

The overall temperature at the junction is given as:

$$T_J = P_{ave}\left(R_{\theta cs} + R_{\theta sa} + R_{\theta jc}\right) + T_A$$

Therefore, the upper limit of the thermal impedance from the sink to the ambient air can be given as:

$$R_{\theta sa} < \frac{T_J - T_a}{P_{arg}} - R_{\theta cs} - R_{\theta jc}$$

The average power dissipated in an active component, P_{avg}, such as that experienced by the SiC power MOSFET selected by the team, is given by the sum of the average conduction losses and average switching losses.

To estimate the "worst case" P_{avg} value experienced by one of the team's selected Power MOSFETs, the worst case values for the conduction and switching losses computed in sections A and C are utilized:

$$P_{\text{ave}} = P_{\text{ave,conduction}} + P_{\text{ave,switching}}$$

Thus, the upper limit of the thermal impedance from the sink to the ambient air can be calculated as mentioned previously. Given this limit, the team sought a heat-sink that stayed below this thermal impedance limit even in the "worst case" thermal impedance of the heat-sink itself. A heat-sink that meets this constraints was identified, specifically the Fischer Elektronik LA 6/100 12 V, [8], which was deemed favorable by the team due to its compact construction, and its ability to retain a thermal impedance less than $R_{\theta sa} = 0.2818$ °C/W, and is relatively simple in cooling operation, utilizing a fan and forced air. Thus, using the upper limit of $R_{\theta sa}$ calculated previously, the total thermal impedance of the packaged inverter, that is, the overall thermal impedance from the junction to the ambient air, in the worst case, can be calculated as:

$$R_{\theta} = R_{\theta jc} + R_{\theta cs} + R_{\theta sa}$$

4 Packaging Arrangement

Throughout the device selection process, design phase of the inverter, and thermal calculations, the physical container for the circuit was considered. The circuit will be mounted on an insulating material and put into aluminum housing. The housing could include small fins on the outer layer of the aluminum box to help with radiating heat as well as the heat sinks and their fans. Other possibilities for packaging could include liquid cooling and further improvements on the forced air ventilation. The overall packaging architecture of the inverter is crucial to its longevity. If not properly cooled and maintained, the devices within will likely burn up, or at least dramatically shorten their lifetime. Choosing equipment that is rated to withstand well above given thermal characteristics is essential and will save money in the long run, as devices will last longer.

5 Drive Circuit and Control

The team utilizes a two-level pulse width modulation (PWM) scheme to drive the inverter, simulating it using the two-level PWM generator available in Simulink. This technique utilizes a sinusoidal saw tooth carrier signal, which is compared against a reference signal *Uref,* also called a modulating signal. When the reference signal value is greater than the carrier, the pulse for the upper switching device is logic high, and the pulse for the lower switching device is logic low [9]. This is summarized in Fig. 4.

A challenge arose in designing the PWM scheme for the inverter regarding how to properly modify the two level PWM reference voltage when the load experienced by the inverter changes. As a solution, the team implemented a three-stage closed-loop control scheme to ensure accurate operation of the PWM scheme regardless of the load (Fig. 5).

The three stages of the team's control scheme can be summarized as follows, which are depicted in Fig. 6:

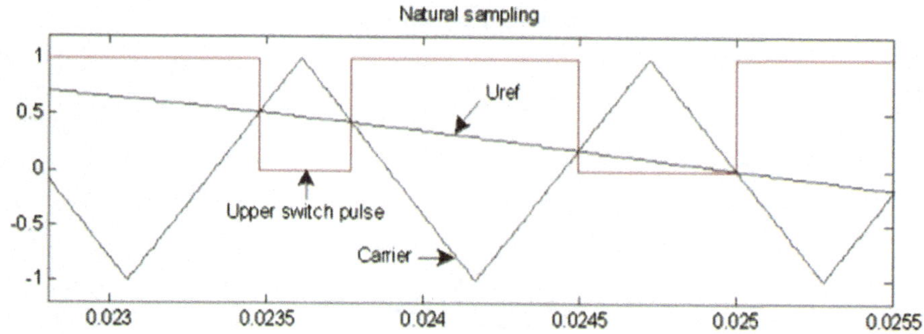

Fig. 4 Two-level PWM example in Natural Sampling mode [9]

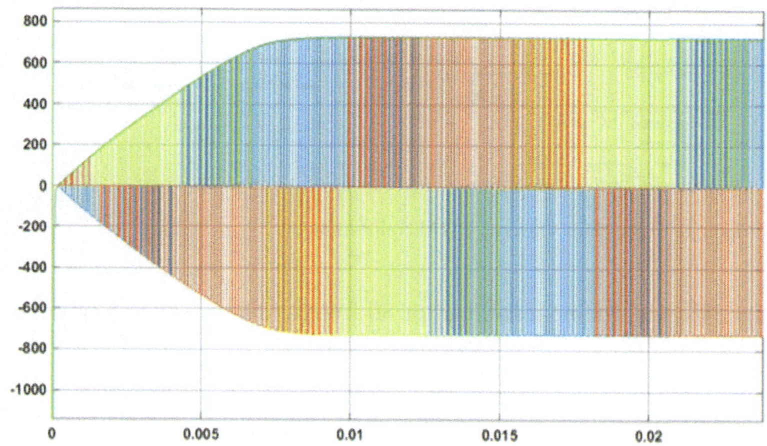

Fig. 5 Pulse width modulated phase-to-phase inverter output voltage waveforms before filtering (Red, Blue, Yellow). Input PV voltage envelope shown in green

1. The measured three-phase output voltage of the inverter V_{abc} is fed into a Phase-Lock-Loop to extract the current value of ω_t.
2. The PLL-calculated value of ω_t and the three-phase output voltage signal V_{abc} are input into a Park Transformation in order to calculate the peak value of the three-phase output voltage [10].
3. The extracted peak value of the three-phase output voltage is subtracted from the desired "reference" peak value of the three-phase output voltage. This difference is fed into a PID controller. The output of the PID controller is multiplied by a "dummy" three-phase waveform with magnitude 1 and 120 degrees apart to form the new three-phase reference voltage for the PWM generator.

The resultant three-phase U_{ref} signal is then fed into the PWM generator, and the above control process is repeated iteratively at every time step. Favorable results were observed with this control scheme, generating the desired output waveforms while within rated load constraints, with an overall favorable total harmonic distortion of

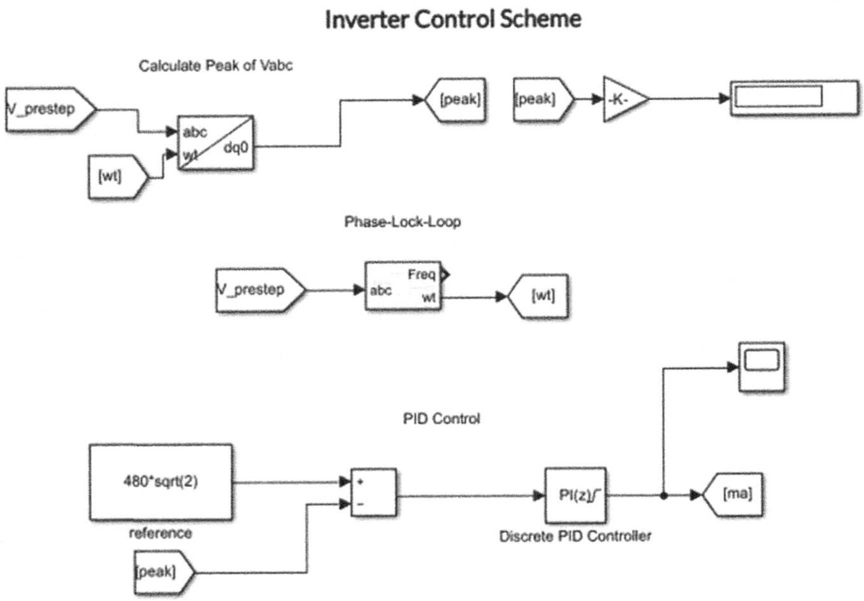

Fig. 6 Topology of the team's inverter PWM control scheme

less than 5%, with the nominal load of 80 kW yielding a THD of approximately 1.8%.

6 Discussion

Some common industry practices used in the design of grid-tie inverters that were not utilized in the team's final model were researched, to be considered for utilization in future work. Chief among these is Maximum Power Point Tracking (MPPT). In short, an MPPT device, which is essentially a microcontroller, attempts to keep the PV panels operating at the point on their i-v curves such that maximum power is generated. This is an optimizing technique that ensures the full capability of the PV array is being utilized. Finding this operation point requires extensive sensor schemes, and an algorithmic controller that increases operating voltage and decreases operating current, or vice versa, and a corresponding closed loop control system that iteratively produces these changes. Additionally, the team determined that, if and when Las Cruces, NM allows net metering, this PV inverter would need the addition of a "**DC Gain Stage**" in order to assist with grid-tie operations.

7 Results and Conclusion

The team's model was simulated for three different loads that could realistically be encountered by the inverter during a year of operation of the Corbett Student Union PV system. The results of the three simulated loads are summarized below in Table 1, and are explained further throughout the section.

Accounting for the single phase nature of the PV source, the total three-phase input power than can be delivered by the portion of the Corbett Center Student Union PV system simulated was found as 119.84 kW. On the rated per-phase load of 80 kW, the measured three-phase power dissipated was simulated as 100.5 kW, Yielding an efficiency of 83.6%. Additionally, the total harmonic distortion per phase was measured as: $\text{THD}_A = 1.733\%$, $\text{THD}_B = 1.760\%$, $\text{THD}_C = 1.756\%$. All of which fall below the grid interconnection standard for THD specified by IEEE Standard 519, which limits harmonic voltage distortion on power systems to 3%.

Figure 7, below, depicts various outputs at the start of the simulation for the per-phase load of 80 kW. Note that the presence of a "**settling time**" for the PID controller produces unacceptable harmonics for approximately nine millisecond during startup time as the PV voltage responds. Future research is needed to determine how to reduce this spike in harmonics.

A "heavy" load of 100 kW **per-phase** was simulated to observe the effects of exceeding the rated per-phase load of 80 kW. From the simulation, the total three-phase input power was found as 135.97 kW. At a per-phase load of 100 kW, the measured three-phase power dissipated was simulated as 112.85 kW, yielding an efficiency of 83%. Additionally, the total harmonic distortion per phase was measured as: $\text{THD}_A = 1.811\%$, $\text{THD}_B = 1.780\%$, $\text{THD}_C = 1.66\%$. Yielding an average THD of 1.76%. The output graphs for this simulation are shown below in Fig. 8.

A "light" load of 40 kW **per-phase** was simulated to observe the effects of a load below the rated per-phase load of 80 kW. A practical example of such a load could be nighttime operation. From the simulation, the total three-phase input power was found to be as 67.7 kW. At a per-phase load of 40 kW, the measured three-phase power dissipated was simulated as 57.15 kW, yielding an efficiency of 84.4%. Additionally, the total harmonic distortion per phase was measured as; $\text{THD}_A = 1.778\%$, $\text{THD}_B = 1.791\%$, $\text{THD}_C = 1.790\%$.Yielding an average THD of 1.78%. The output graphs for this simulation are shown below in Fig. 9.

Some remarks regarding the results of the simulations can be made. As expected, an increase in load resulted in a proportional increase in the magnitude of drawn output current. In regards to output voltage, as the load approached the rated values

Table 1 Summary of simulation results for all loads

Load per phase	$P_{i,3\varnothing}$ (kW)	$P_{o,3\varnothing}$ (kW)	η (%)	THD_{avg} (%)
40 kW	67.72	57.15	84.39	1.78
80 kW (rated)	119.84	100.5	83.86	1.75
100 kW	135.97	112.85	82.99	1.76

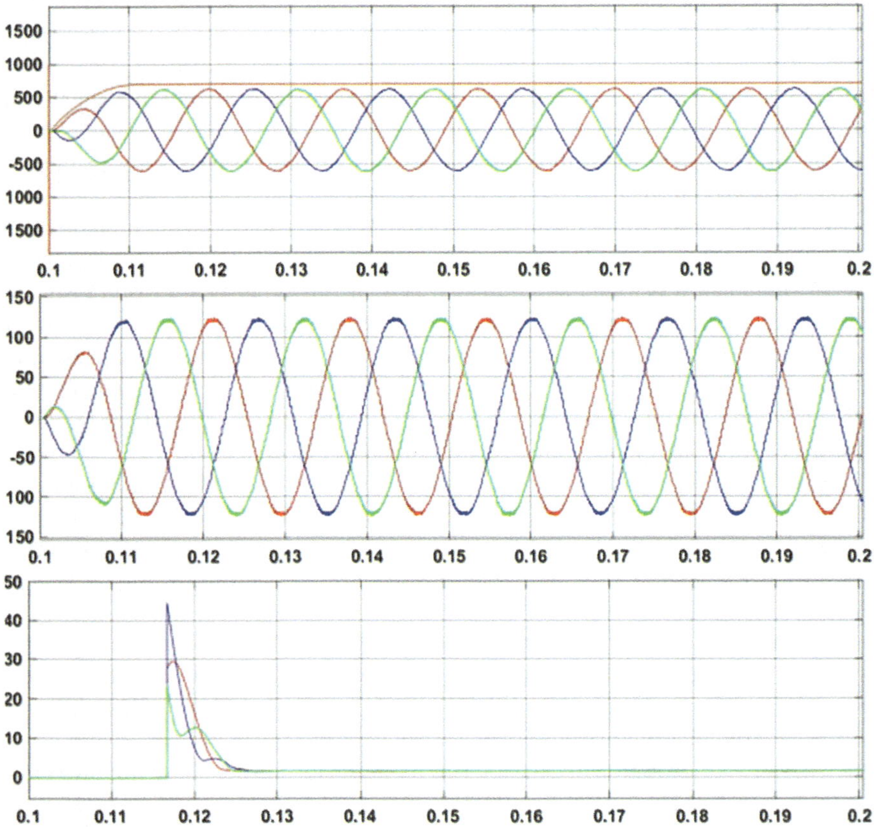

Fig. 7 80 kW per-phase loading results. All plots versus time. From top bottom: 1. Phase-to-phase output voltages of the inverter after step-up transformer (Red, Blue, Green) and the PV voltage (Orange) 2. THD values of the phase-to-phase output voltages

of the panels, some voltage regulation challenges were encountered. Specifically, the magnitude of the output voltage began to very slightly decrease as the load approached the rated value, and was unable to be perfectly regulated to grid voltage by the control system. The team hypothesized that this was likely due to non-linear losses incurred due to the filter components, or a need for improvements in the control system. The importance of this was further verified by additional research by the team that found, according to [11], that modern "high-quality" sine wave PV inverters typically possess efficiencies of around 90–95%. Inverters deemed of lower quality were said to have efficiencies of approximately 75–85% [11].

Preliminary calculations were performed to attempt to quantify the unexplained non-linear losses observed in the simulation. Using the calculated efficiencies for the 80 kW and 100 kW per phase loads, η_{80} and η_{100}, the non-linear losses due to the filter and reactive components of the H-bridge were attempted to be calculated as:

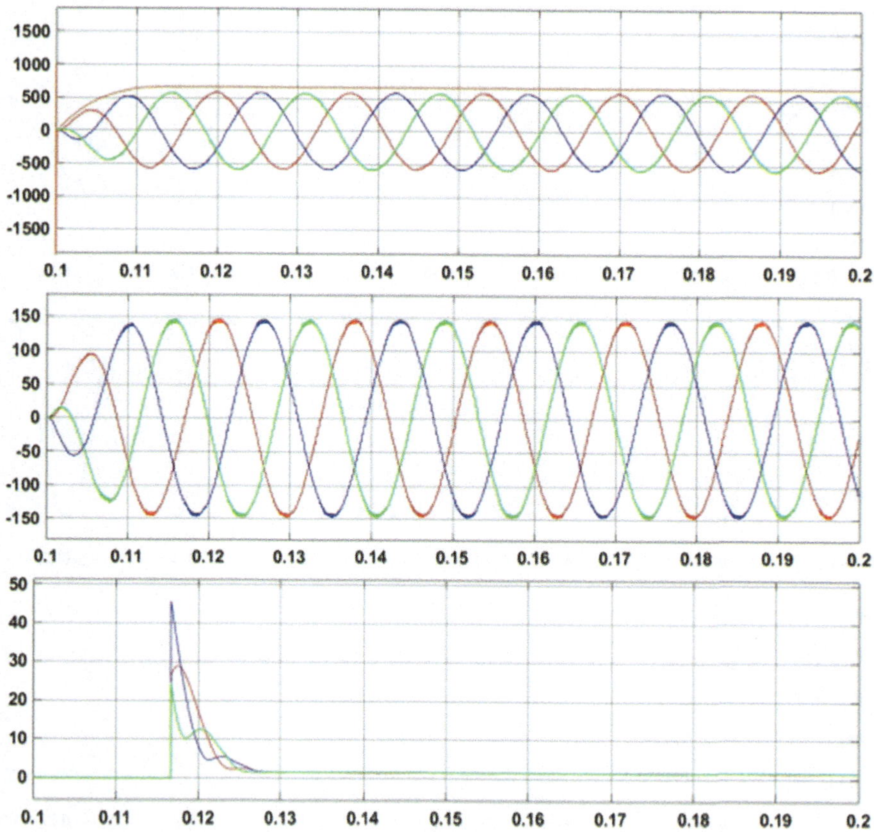

Fig. 8 100 kW per-phase load results. All plots versus time. From top bottom: 1. Phase-to-phase output voltages of the inverter after step-up transformer (Red, Blue, Green) and the PV voltage (Orange) 2. THD values of the phase-to-phase output voltages

$$\frac{\text{loss}}{\text{dec}} = \frac{\eta_{80} - \eta_{100}}{\text{dec}^{-1}} \cdot 100, \text{ and dec} = \log\left(\frac{80}{100}\right)$$

In conclusion, this project was instrumental in introducing the team to the iterative process of designing power electronic systems. Not only was it a phenomenal learning experience, but it also exposed the team to the breadth and depth of design considerations and problems that are encountered throughout the design and modeling of a grid-tie photovoltaic inverter.

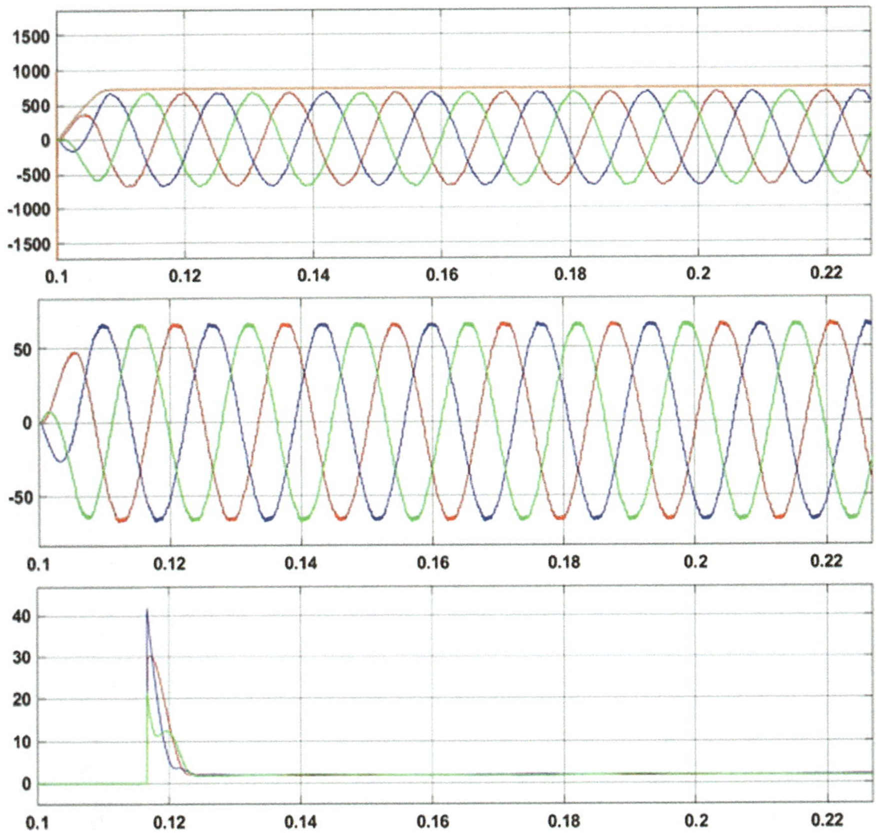

Fig. 9 40 kW per-phase load results. All plots versus time. From top bottom: 1. Phase-to-phase output voltages of the inverter after step-up transformer (Red, Blue, Green) and the PV voltage (Orange) 2. Output phase currents of the inverter 3. THD values of the phase-to-phase output voltages

References

1. M. Slovick, Working with SiC MOSFETS: Challenges and Design Recommendations, digikey.com, Sept. 27, 2017
2. M. Vujacic, M. Hammami, M. Srndovic, G. Grandi, Analysis of dc-Link Voltage Switching Ripple in Three-Phase PWM Inverters (2018)
3. Silicon Carbide Power MOSFET C2M. C2M0025120D. Mouser (2015)
4. Super High Power BiFacial Poly Perc Module. CS32-420 PB-AG. Canadian Solar Inc (2019)
5. M.H. Rashid, *Power Electronics: Devices, Circuits, and Applications* (Pearson, Boston, MA, 2014)
6. N. Guler, E. Irmak, MPPT based model predictive control of grid connected inverter for PV systems, in *2019 8th International Conference on Renewable Energy Research and Applications (ICRERA)*, (Brasov, Romania, 2019)
7. Infineon Automotive Power Staff, MOSFET Power Losses Calculation Using the Datasheet Parameters, Infineon (2013)

8. Cooling aggregates with axial fan. LA 6/100 12 V. Fischer Elektronik
9. PWM Generator (2-Level), Generate pulses for PWM-controlled 2-level converter-Simulink
10. P. Krause, O. Wasynczuk, S.D. Sudhoff, S. Pekarek, *Analysis of Electric Machinery and Drive Systems* (Wiley-IEEE Press, Piscataway, NJ, 2013)
11. 6.5. Efficiency of Inverters, 6.5. Efficiency of Invertersl EME 812: Utility Solar Power and Concentration

Smart Power Management System for Charging Plug-in Hybrid/Electric Vehicles Using Solar PV for Software Technology Park

Kartik Virmani, Y. Raja Sekhar, Akshat H. Mutta, Tarun Sharma, and Naushad Ali

Nomenclature

I	Output current
I_{ph}	Photocurrent
I_d	Diode Current
I_0	Saturation Current
V	Voltage Imposed
V_T	Thermal Voltage
N_s	Number of PV cells connected in Series
A	Ideality Factor
RP	Parallel Leakage Resistance
S	Series Resistance
k	Boltzmann Constant
Tc	Critical Temperature of Diode
IP	Parallel Leak Current
Iph, ref	Photocurrent (A) at STC
Tc, ref	Cell temperature at STC = 25 + 273 = 298 K, Isc, provided by the manufacturer
P_{mp}	Power at the maximum PowerPoint
G	Solar Irradiation
q	Electron Charge
B_{sc}	Battery Storage Capacity
N_{ccd}	Battery Storage Capacity
D_d	Depth of Discharge of Battery
n_b	Temperature Derate Factor

K. Virmani · Y. Raja Sekhar (✉) · A. H. Mutta · T. Sharma · N. Ali
School of Mechanical Engineering, Vellore Institute of Technology, Vellore, Tamil nadu 632014, India
e-mail: sekhar.rajay@gmail.com

V. K. Singh et al. (eds.), *Renewable Energy and Future Power Systems*, Energy Systems in Electrical Engineering, https://doi.org/10.1007/978-981-33-6753-1_7

n_l	Rate Factor
n	Day Number
EOT	Equation of Time
LAT	Local Apparent Time
α	Altitude Angle
γs	Solar Azimuth Angle
D	Distance between panels
H	Height of the panels
CF	Capacity Factor

1 Introduction

Energy from Photovoltaic (PV) cells is a clean form of power and disassociates any ozone-depleting substance during its period of operation. PV has an incredible possibility in various applications on account of the decrease in cost with time and improvement in its effectiveness due to enhancements in material science [1]. Besides these, the photovoltaic system design at any location is additionally done based on industrial standard principles. Usually, the installation of standalone photovoltaic frameworks need not require transmission and distribution equipment for satisfying the client needs with pertinent quality [2, 3]. Further, various open access solar radiation information provided by satellite has both high spatial and transient goals. Software simulation of PV systems will make one understand the merit and demerits in terms of technical and economic viability.

In the current scenario, the Indian government is encouraging research on developing smart transport for rural and urban communities. Also, Make-in-India policy is encouraging start-ups to come up with innovative ideas that create commercialization. One of the solutions for smart transport is electric vehicles where there is an immediate necessity to address the potential issues brought about by the rise of Plug-in Hybrid Electric Vehicle (PHEV)/Plug-in Electric Vehicle (PEV) charging requirements. Micro-controller based smart charging system implementation; reduce the burden on charging devices in terms of hierarchical control [6]. However, the designer needs to consider, the battery specifications, along with charging specifications to come with an optimized control system. For the future, the system should be capable of re-configuring the charging layout to ascertain expansion due to the increase in vehicles with time.

In the case of PHEV, there will be a combination of both the combustion engine and electric motor connected in either series or parallel based on the power requirement. In a PHEV, the engine drives the electric motor through a dynamo, which in turn provides the transmission. One of the real-time problem encountered in PHEV is that due to fuel shortage at times, the motor is driven by storage batteries. However, there is a limitation on the range since, the battery State of Charge (SoC) is less.

Hence, to overcome this issue, smart charging stations will assure that energy to charge batteries is accessible at regular intervals.

This chapter discusses electric vehicle charging stations utilizing PV panels at parking lots in IT parks or Technological parks. The proposed framework is intended for use in working environments to charge electric vehicles of the staff at parking lots during the day. The purpose is to boost the utilization of PV vitality for charging EV on a large scale in contrast to the existing grid-based network. Research gaps related to this concept in current literature lead to the following objectives:

1. Develop optimum feed-back control based charging system to enhance the utilization of PV as well as EV in India
2. A method to arrive at the possible PV system power rating based on metrological conditions prevailing at different states and reduce system oversize using MATLAB Simulink platform.
3. The control sequence adopted for this study can be easily adapted to other global conditions by changing the meteorological parameters to arrive at any scale of charging stations for PHEV/EV.

2 Location

The location chosen for setting up the solar charging station for electric cars is an information technology (IT) park situated at Chennai. The latitude and longitude of the location are 12.98° N, 80.24° E. Chennai region has an abundant amount of sunshine, making it an optimum location for our charging station. The IT park named TIDEL park is an ISO 9001/14001 company and one of the largest in Asia set up in the year 2000. Two underground parking spaces and one outdoor parking area having space for 4,000 two-wheelers and 1,200 cars is available at the park. The outdoor parking space was most suited for setting up our solar charging station.

3 Determination of Charger Specifications

The charging station would be equipped with level one DC charging ports with a charging supply of 15 kW to charge the vehicles. The supply voltage and current are 240 V and 60A, respectively. At IT parks, the cars are generally idle for about 8-10 h depending on the working hours, and that's why level one charging ports are selected for charging the cars. Data was collected on the battery capacity of various electric cars available in India in addition to some of the upcoming car launches to determine the DC supply load. From the data, it was observed that one vehicle would require up to 75KWh of electric supply to charge the vehicle from a 0% SoC to 100% SoC. The charging station planned would be able to charge 150 cars at a time. It would take approximately five hours to charge the vehicle at 15 kW supply. The methodology adopted in this study can be generalized to any location around the world, with just altering a few parameters that will be discussed later in this study.

4 Power Generated from the PV Modules

Internal losses in the PV modules are neglected in the equations. Hence as the solution, a diode in parallel with the photon generated current source[11]. The current at the output is calculated by applying Kirchhoff's law:

$$I = I_{ph} - I_d \tag{1}$$

$$I_d = I_o\left[exp\left(\frac{V}{A.N_s.V_t}\right) - 1\right] \tag{2}$$

In reality, the effect of series and parallel resistance, R_s & R_p respectively cannot be neglected considering their importance when it comes to the efficiency of the PV module. Considering the impact of R_s & R_p in the circuit, the equation modifies to (Fig. 1):

$$I_d = I_o\left[exp\left(\frac{V + I.R_s}{a}\right) - 1\right] \tag{3}$$

where

$$a = \frac{N_s.A.k.T_c}{q} = N_s.A.V_T \tag{3a}$$

Now,

$$I = I_{ph} - I_d - I_p \tag{3b}$$

where parallel leak current is considered.

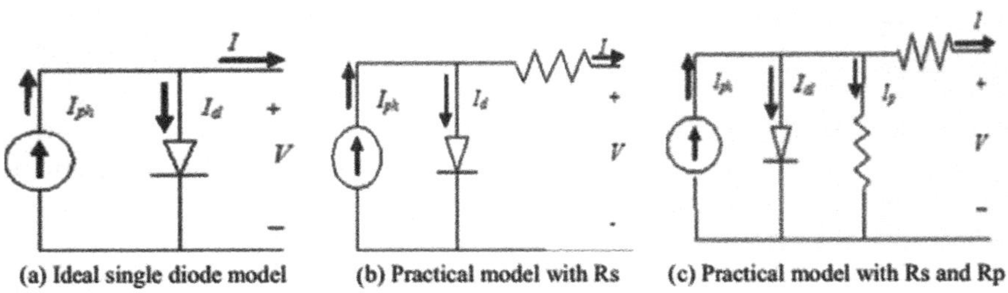

Fig. 1 Equivalent Circuit of Photovoltaic cell in a Solar device [11]

$$I = I_{ph} - I_o\left[exp\left(\frac{V + I.R_s}{a}\right) - 1\right] - \frac{V + I.R_S}{R_p} \quad (4)$$

5 Module Selection

The module model selected for the study is Somera Grand Ultima Max *SOMERA VSMH.75.AAA.05*, manufactured by, M/s. Vikram Solar Cell limited, India. The module has monocrystalline 150 solar PV cells together delivering about 405-425 Watts.

The harvested current gathered by the PV module formation(i.e. the array) relies on the voltage generated by PV array, the amount of solar irradiance on the array, wind rate of the region, and temperature of the surroundings(i.e. the ambient temperature) [13]. The module currents I_{module} can be obtained by adding up the currents passing across each line (i.e., Ns sets) and power from the module is given by Eq. (5).

$$P_{module} = (I_{module})(V_{module}) \quad (5)$$

6 Determination of Parameters

It was examined that I_{ph}, I_o, R_s, R_p, and *ideality factor* are the known limitations to any Solar Photovoltaic System that are governed by the incident solar radiation and the cell temperature [19].

6.1 Determination of Iph

$$I = I_{ph,ref} - I_{o,ref}\left[\exp\left(\frac{V}{a_{ref}}\right) - 1\right] \quad (6)$$

$$I_{sc,ref} = I_{ph,ref} - I_{o,ref}\left[\exp\left(\frac{0}{a_{ref}}\right) - 1\right] \quad (7)$$

G is the cell Irradiance (W/m^2), G_{ref}: Irradiance at STC = 1000 W/m^2, $\Delta T = (T_C - T_{ref})$

6.2 Determination of I_o

$$I_{o,ref} = I_{sc,ref}\exp\left(\frac{-V_{oc,ref}}{a}\right) \tag{8}$$

The reverse saturation current is given by

$$I_o = I_{o,ref}\left(\frac{T_c}{T_{c,ref}}\right)^3 exp\left[\left(\frac{q\varepsilon_G}{A.K}\right)\left(\frac{1}{T_{c,ref}} - \frac{1}{T_c}\right)\right] \tag{9}$$

Substituting Eq. (8) for I_0,

$$I_o = I_{sc,ref}\exp\left(\frac{-V_{oc,ref}}{a}\right)(\frac{T_c}{T_{c,ref}})^3 \times exp\left[\left(\frac{q\varepsilon_G}{A.K}\right)\left(\frac{1}{T_{c,ref}} - \frac{1}{T_c}\right)\right] \tag{10}$$

6.3 Determination of R_s and R_p

A Closed-loop response is simulated, i.e. used to get the value Rs and Rp, Simulation is run until the same result for Power was Pmp, and Pmp Experimental was obtained.

$$I_{mp,ref} = {}^{P_{mp,ref}}\!/\!{}_{V_{mp,ref}} = {}^{P_{mp,ref}}\!/\!{}_{V_{mp,ref}} \tag{11}$$

$$I_{mp} = I_{ph,ref} - I_{o,ref}\left[exp\left(\frac{V_{mp,ref} + I_{mp,ref}.R_s}{a}\right) - 1\right] - \frac{V_{mp,ref} + R_s.I_{mp,ref}}{R_p} \tag{12}$$

$$R_p = \frac{V_{mp,ref} + I_{mp,ref}.R_s}{I_{sc,ref} - I_{sc,ref}\left(exp\left[\frac{V_{mp,ref}+R_s.I_{mp,ref}-oc,ref}{a}\right]\right) + I_{sc,ref}\left[exp\left(\frac{-V_{oc,ref}}{a}\right)\right] - (P_{max,ex}/V_{mp,ref})} \tag{13}$$

6.4 Methodology Used to Evaluate the Parameters (MATLAB and SIMULINK)

It's important to note that the Simulink model, as shown in Fig. 2 evaluate the photocurrent and short circuit when there is one cell in series. Whereas in the final model using subsystems, these values evaluate the output current, voltage and the number of cells required in series and parallel that will further help in estimating the price of the Solar cells [18]. Thus, this allows us for optimal simulation of current and voltage as required by the load throughout the year and further help in battery

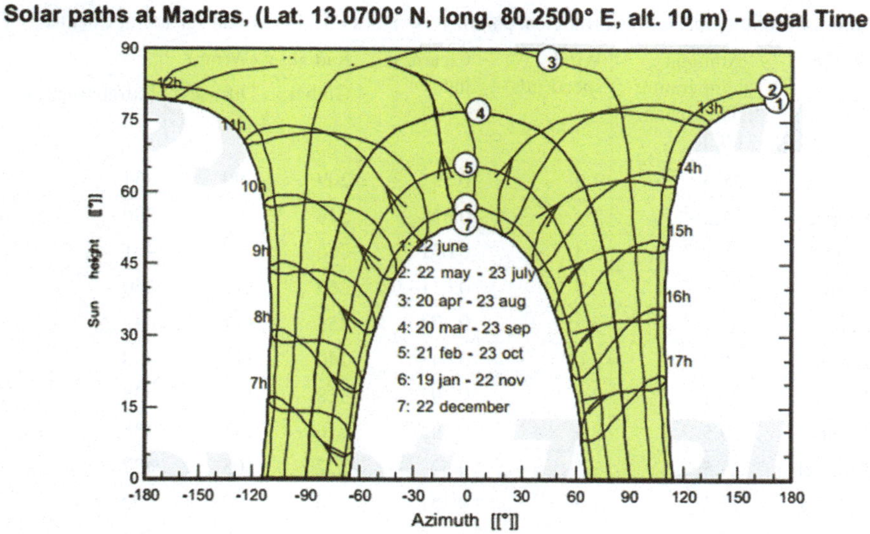

Fig. 2 Sun Path Diagram

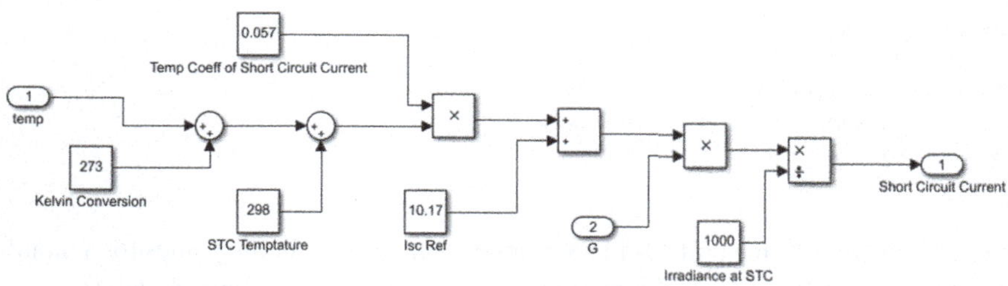

Fig. 3 Simulink Model to determine I_{ph}

sizing and proper charge controller selection. By model simulation, the resistance value is **0.9632 ohms**. For the series resistance due to the diode effect and parallel leakage, resistance is about **520 ohms** [19] (Fig. 3).

6.5 Evaluation of Number of Solar Cells Required

After knowing all the required parameters above as Simulink model, the final model was created where the above models acted as the subsystems as like the sub-branches to the main tree. The value of the Irradiation, G, and the temperature was taken up from Table 1, and other essential parameters and constants were directly taken from the manufacturer's datasheet. Signal Builder used in the model was run for 12 units of time until obtained the data corresponding to each month and the year around

Table 1 Monthly mean Meteorological parameters values on a horizontal surface at Chennai

Month	Ambient temperature daily mean (°C)	Wind speed(m/s)	Clearness index	Radiation (W/m^2)		
				Global	Diffuse	Extraterrestrial
January	25.9	1.9	0.59	209	90	354
February	26.2	2	0.628	244	87	390
March	28.1	2.4	0.631	266	95	421
April	29.7	2.9	0.611	269	111	440
May	31.7	3.5	0.574	253	114	442
June	30.8	3.6	0.548	240	118	438
July	30.4	3.1	0.497	218	117	438
August	29.5	3.1	0.493	216	124	438
September	28.5	2.5	0.52	222	105	427
October	27.3	1.5	0.451	180	99	399
November	25.6	1.5	0.455	165	97	363
December	24.1	1.9	0.467	160	87	342
Annual	28.2	2.5	0.539	220	104	408

Latitude 13.07°N
Longitude 80.25°E
Legal Time zone UT+5.5
Source MeteoNorm 7.2 station
PV Syst 6.86

average, where the time units 0–11 each corresponding to specific values for a month and finally the 12th time step depicting the year around value (Figs. 4 and 5).

The Signal builder, as shown in Fig. 6, imports data for the Daily Mean Temperature, Irradiation, and the Mean Sunshine hours around the month, as stated in Table 1.

7 Battery Selection

To create a sustainable PV system for the presumed load, the capacity of the battery being used in addition to charge controllers had to be taken into account, and it must be calculated. Usually, the approach for battery design is that in case of larger capacity requirement (i.e. larger Ah), "lower-voltage, higher Ah batteries" is preferred. Since, numerous batteries having lower voltage can be connected in series such that lower number of sets of batteries can be in parallel [12]. Equation (14) give the batteries to be connected in series.

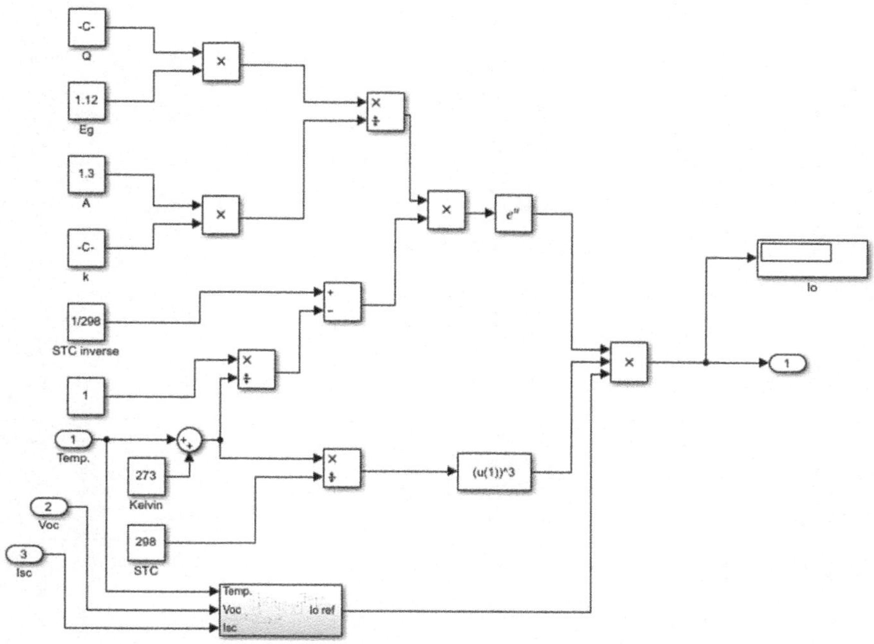

Fig. 4 Simulink Model to determine I_o

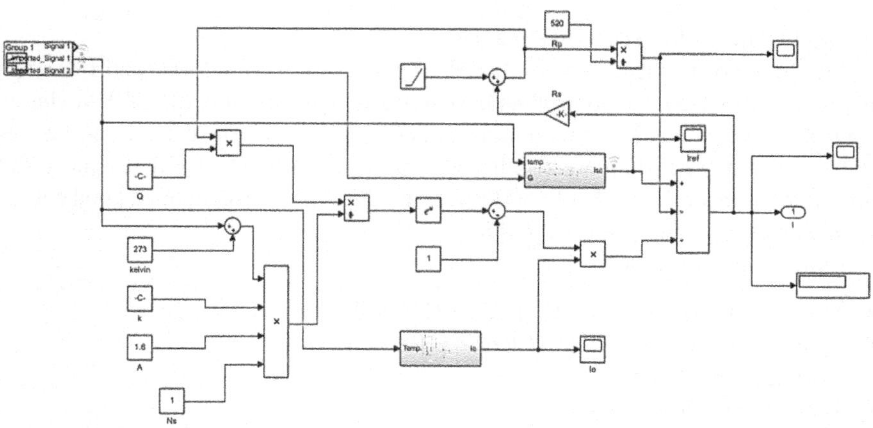

Fig. 5 Simulink Model to determine resistances

$$B_{sc} = \frac{N_{ccd} \times L}{\eta_B \times D_d \times \eta_I} \tag{14}$$

For this study,
Battery Rate Factor = 1.25; Temperature de-rate Factor = 0.95
Days of Autonomy = 4

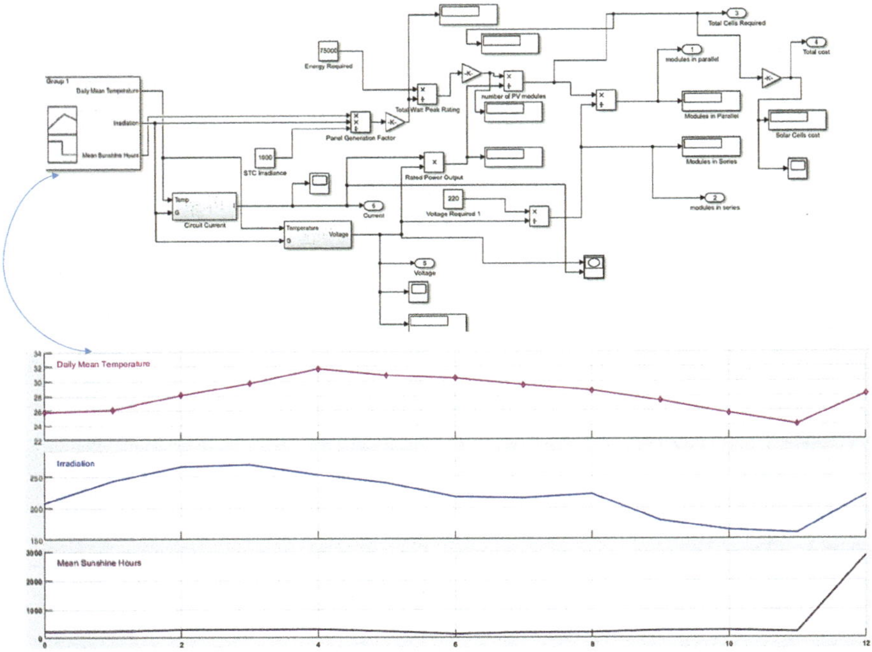

Fig. 6 Signal Builder (Input) Solar Radiation and Ambient Temperature Profile for Chennai

Total Hours of Battery usage = 8 h

For the analysis, the MPPT solar charge controller rated for 240 V and 100 amps is chosen. MPPT charge controllers have a higher current rating than PWM charge controllers which makes them optimum for our application. The battery chosen was a lead-acid battery by luminous with a capacity of 150Ah and 12 V. The battery had 60 months warranty and life of 1500 cycles at 80%DoD making them a good choice for the charging station (Figs. 7 and 8).

Battery Specification:

- Model: Solar Battery 150 Ah–LPTT12150H (Luminous Solar Battery)
- Rating: 150 Ah Capacity as stated @ C10, 12 V
- Warranty—60 Months
- Tubular plate Technology
- DOD Criteria:
  - 80% DOD with 1500 cycles
  - 20% DOD with 2000 cycles

The battery delivers a longer life with a self-discharge rate of 3% per month at STC.

Battery Characteristics assumption that Higher A-h efficiency > 90%; Watt Hour efficiency > 80% (Table 2).

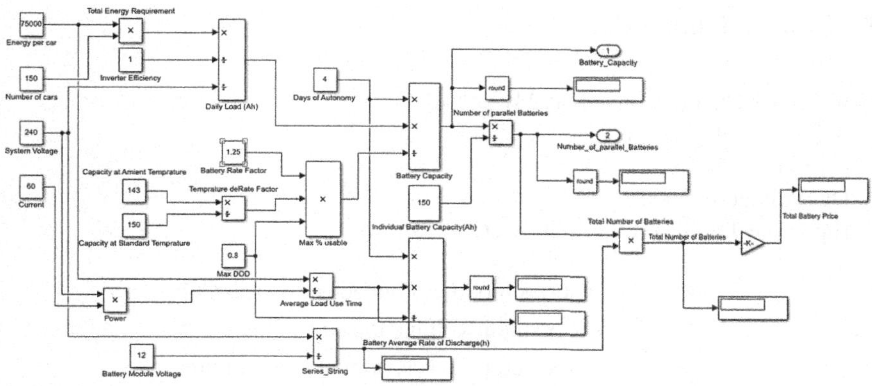

Fig. 7 Simulink Model including Battery Sizing

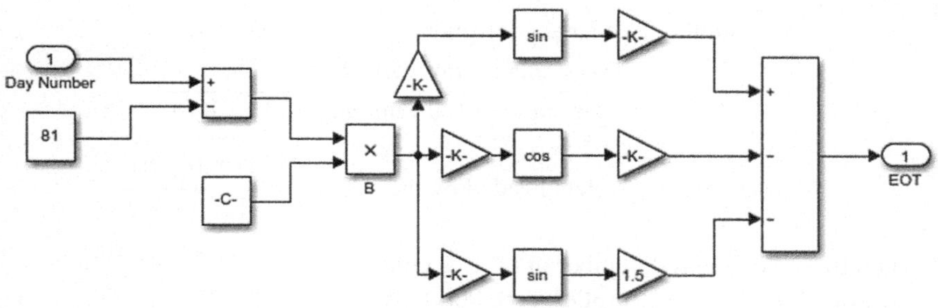

Fig. 8 Simulink Model to calculate EOT

Table 2 The ideal number of PV modules and batteries in the array configuration

Month	Number of Modules			Number of batteries		
	Series	Parallel	Total	Series	Parallel	Total
January	5	20931	104655	15	1290	19350
February	5	14863	74315	15	1290	19350
March	5	10279	51395	15	1290	19350
April	5	9944	49270	15	1290	19350
May	5	10965	54825	15	1290	19350
June	5	15682	78410	18	1300	19350
July	5	31220	156100	18	1300	23400
August	5	24014	120070	18	1300	23400
September	5	22071	110355	20	1311	23400
October	5	22516	112580	20	1311	26220
November	5	30023	150115	20	1311	26220
December	5	39626	198130	20	1311	26220

8 Charge Controller

MPPT Solar Charge Controller, 24v to 240v

A Product of TESLA

Current = 100 A

DC Voltage = 1000 V

Equations Used for Formulation are as follows

$$\begin{aligned}&\text{Max no. of series modules connected to each CC}\\&=\frac{\text{Max. input voltage the cc can take}}{\text{Voc corrected to lowest annual temperature}}\\&=\frac{240v}{230v}=1.04347\end{aligned}$$

$$\begin{aligned}&\text{The number of charge controllers needed by the system}\\&=\frac{\text{max.array size (Wp)}}{\text{min.battery voltage*current output of each CC}}\end{aligned}$$

$$\text{No.of charge controllers needed by the system}=\frac{11250*10^3}{12v*60A}$$

Therefore, the required number of Charge controllers = 15625.

Cost of Each Charge Controller = 10,000 INR

Total Cost of CC's = 15625 * 10,000 = 156.25 million INR

9 Row Spacing and Sizing

The equations below can be used to calculate the grid spacing for the proposed PV system [14]. It uses the energy calculation method as that for annual shading, and net energy yield model as proposed for the PV system, and an economic proposal, i.e. dependent on the systems investment cost [14]. Thus, the method displayed estimates the several engineering and economic criterion that may influence the ideal array spacing. The optimum model of row spacing was recognized by considering the hour angle from nine in the morning to five in the evening considering a typical job with an 8-h window of sunlight when the Solar Charging Station will be at the peak usage for Chennai.

The formulation was done as follows

For declination angle Eq. (15) is used

$$\delta=23.45\sin\left[\frac{360}{365}(284+n)\right]\tag{15}$$

The Equation Time

$$EOT = 9.87\sin(2B) - 7.53\cos(B) - 1.5\sin(B) \tag{15a}$$

where, B = (360/364) *(n−81)

Then a subsystem was designed that took EOT, Watch Time and Longitude Angle as input and the output thus obtained was Hour angle and LAT.

$$LAT(SolarTime) = WatchTIme - 4 \times (Long_{Std} - Long_{Loc}) + EOT \tag{16}$$

$$HourAngle(\omega) = 15 * (LAT - 12) \tag{17}$$

Further, it was required to determine the altitude angle of the sun and the azimuth angle at the time of the day so that the max probable position of the sun could be obtained and the max row spacing could be performed to prevent the problem of shading year around. The problem was to figure out parameters of the row spacing that are h, D & D'. such that maximized set of values such that implementing a dual-axis control system the shading would never be a problem (Figs. 9, 10 and 11).

So, to reach the optimal set of parameters, a SIMULINK model was framed, and the result was simulated to find the day number of the average days of every month.

The altitude angle was obtained from the following relation

$$\sin\alpha = \sin\delta \times \sin\emptyset + \cos\delta \times \cos\emptyset \times \cos\omega \tag{18}$$

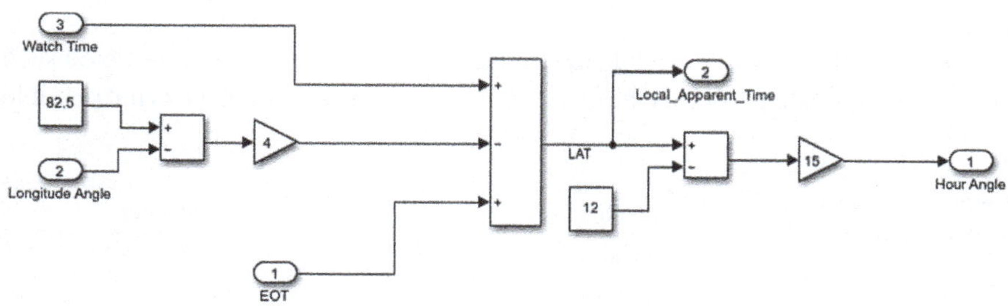

Fig. 9 Simulink Model to determine hour Angle & LAT

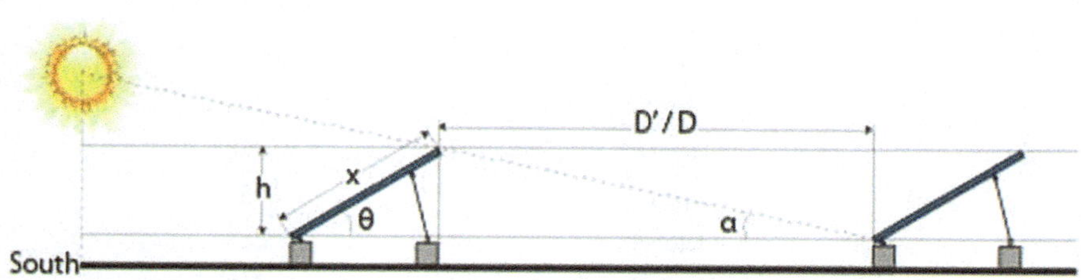

Fig. 10 Sun Altitude Angle to determine the distance between rows in a PV array [23]

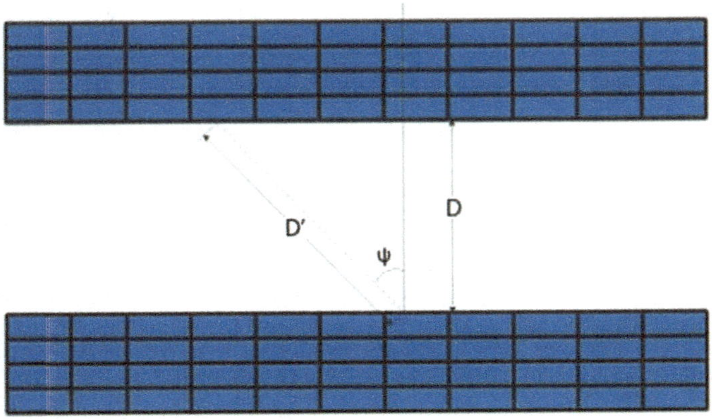

Fig. 11 Distance between the PV module using Sun Azimuth Angle [23]

The azimuth angle was obtained from the following relation

$$\cos \gamma_s = \frac{\cos \theta_z \times \sin \emptyset - \sin \delta}{\sin \theta_z \times \cos \emptyset} \tag{19}$$

The Height of the Solar Panel is given by $h = x \times \sin \theta$, for tilted arrays (Table 3). To calculate the distance,

$$D' = \frac{h}{\tan \alpha}; D = D' \times \cos \psi \tag{20}$$

The tilt angle of the PV module (α) for the concerned scenario is considered such that it is always facing towards the equator. However, it is known that it is not possible

Table 3 Average days of each month in a year [22]

Month	Date	Day number
January	17	17
February	16	47
March	16	75
April	15	105
May	15	135
June	11	162
July	17	198
August	16	228
September	15	258
October	15	288
November	14	318
December	10	344

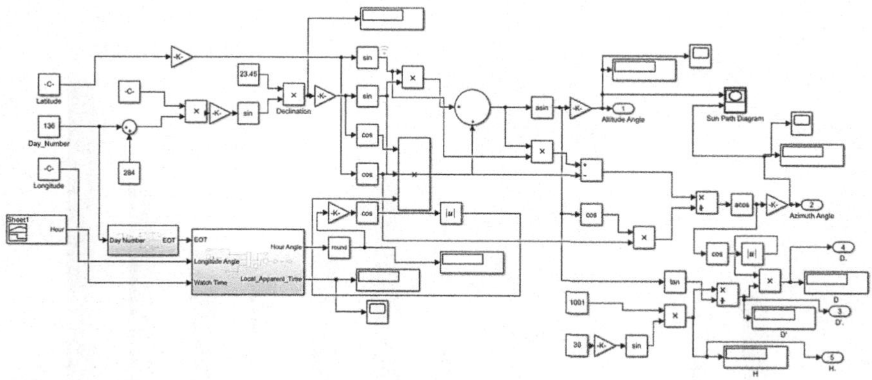

Fig. 12 Simulink Model to calculate row spacing in the PV array

to fit the correct orientation depending on the terrain available the land coverage and the size of the module also may vary depending on the manufacturer; therefore, a solution for that is provided by considering a prescribed variation of $-15°$ to $15°$ to assimilate orientation, i.e. compatible with the available rooftop and land area in which the PV modules are to be arranged [9].

Now to study and analyse the row spacing parameters across a region, a quick and reliable method for the calculating all the parameters that include the angle of tilt and ground height for each row of modules is necessary. The successive distance has to be calculated such that the problem of shading on module for each successive module is never encountered. Daily on each hour the module orientation value change to assure modules trace sun path to receive perpendicular rays incident at the solar panel with a tolerance of $-15°$ to $15°$ during which all the cars remain parked under idle condition every day [9]. As stated earlier, MATLAB and Simulink have been used, all the necessary equations [9, 15] for the determination of module arrangement has been encoded in the Simulink Model as shown in Fig. 12 (See Figs. 13, 14, 15 and 16).

Thus, Figs. 17, 18 and 19 show the results obtained from the analysis. Further, a fact had to be enlightened here that to predict the maximum to minimum land coverage not only depends on the number of modules required but also their arrangement. Therefore, to reach a particular solution to design and economic criterion of the project the maximum row spacing has to be predicted, that is only possible when we run the simulation for that day in the year on which the sun is at the lowest altitude and, i.e. the Winter solstice on 21st December. The results obtained from the simulation (Fig. 12) are summarized in Table 4.

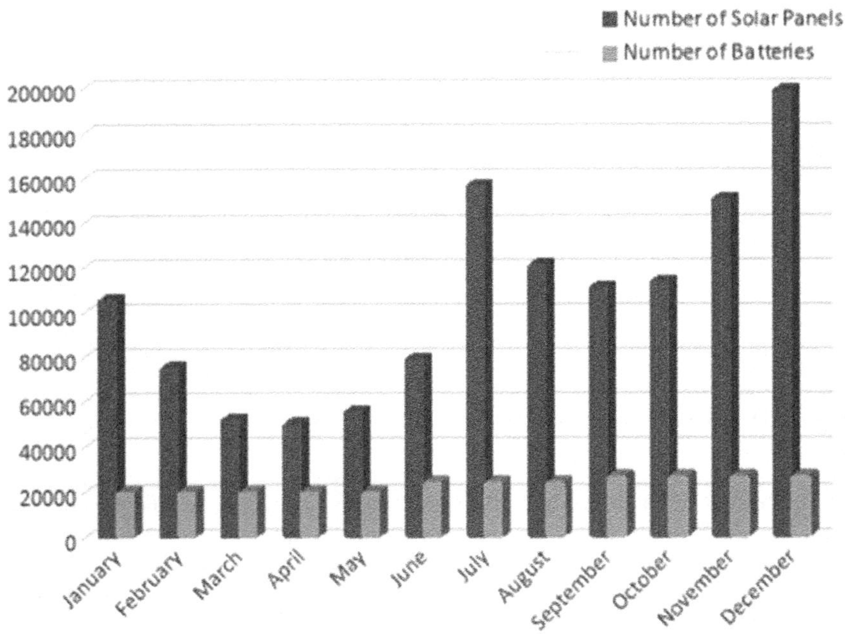

Fig. 13 The required quantity of Solar PV units and batteries for each month

10 Simulated Results and Discussion

Therefore, from the result above it can be foreseen that the station is suitable to handle the least load demand for charging 150 vehicles, i.e. 11250 KW/h at a time, with the peak obtained power in July, but can be successfully operational to a load of 30000 KW/h all the year-round without any issue.

The Figs. 17, 18 and 19 shown above state the variation in row spacing parameters over the watch time from nine in the morning to five in the evening, i.e. a typical 8-h job of an employer in which time car remains idle and can charge simultaneously.

The graphs provide actual data for the variation in the row spacing parameters and simultaneously displaying the changing solar angles on the Mean day of the month.

Note, that only three random sample months graphs are shown in this analysis; however, the Simulink model 7 with day number taken from Table 3 is capable of generating numerical data for every month.

Considering the least case criteria for row spacing, that will occur at Winter Solstice on 21st December.

Hence, the corresponding Day Number is 355.

Furthermore, the parameter variation was also observed on the Summer Solstice on 21st June (Figs. 20 and 21).

Day Number as on 21st June is 172.

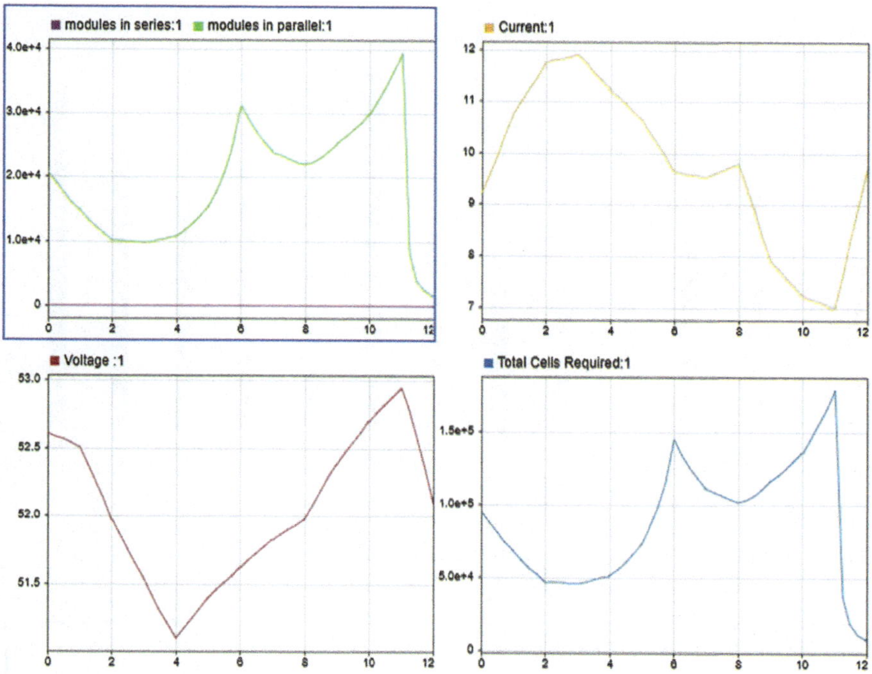

Fig. 14 Number of Cells required in series and parallel throughout the year Plot (1,1), Current variation through each module throughout the year Plot (1,2), Voltage Variation throughout the year from each module Plot (2,1), Total Cells Required Plot (2,2)

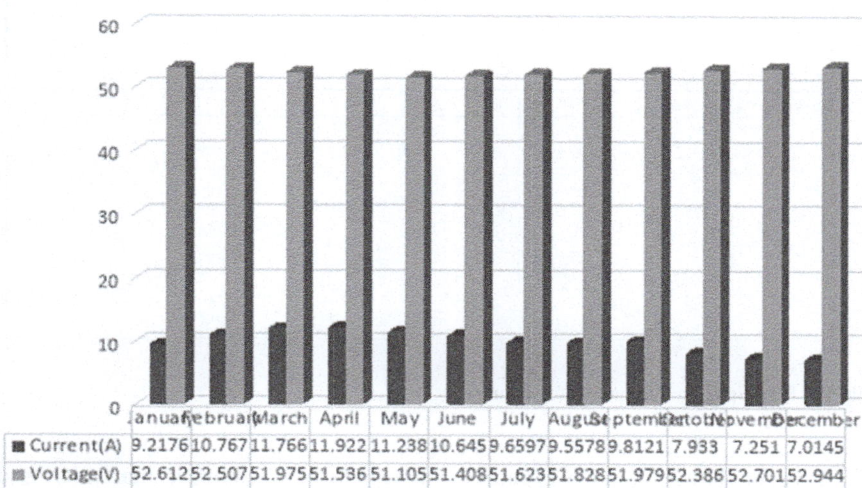

Fig. 15 Monthly Current Output and Voltage from each module

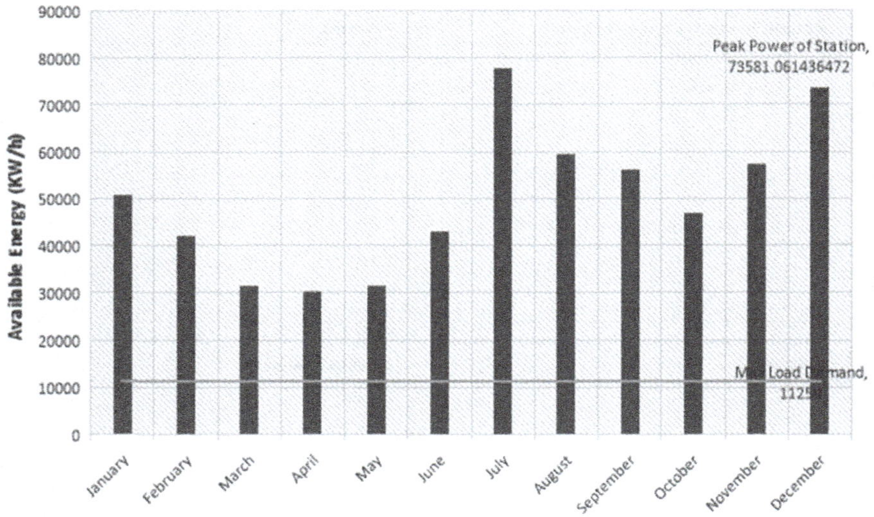

Fig. 16 Variation of Total Energy & Least Load Demand

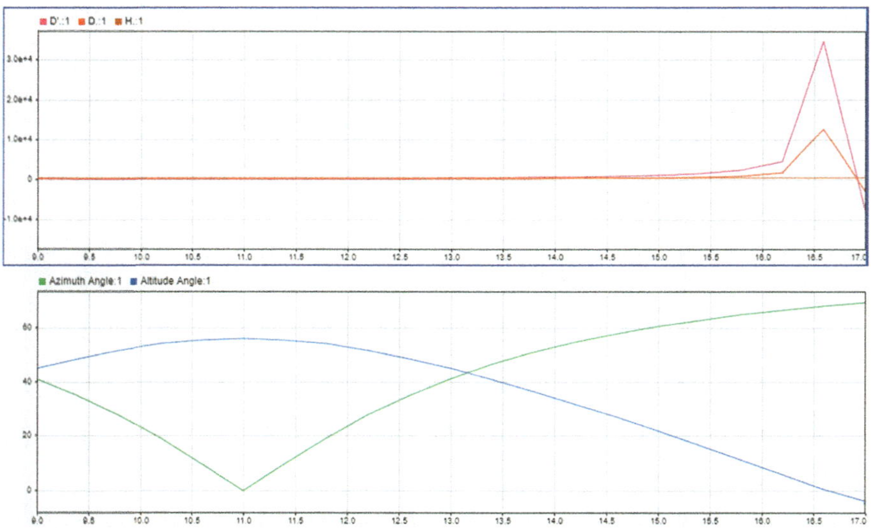

Fig. 17 Row Spacing Parameters for January

Hence, the Simulink model is entirely viable to design a compact smart microgrid and perform the simulation to calculate the number of modules and battery required for setting up the Solar PV charging station at any location. The only input parameters required are the latitude and the longitude that will provide with the environmental parameters of the selected site which can be extracted from the software like PV Syst and PV Sol of the desired location. Also, the average estimate of the total number of

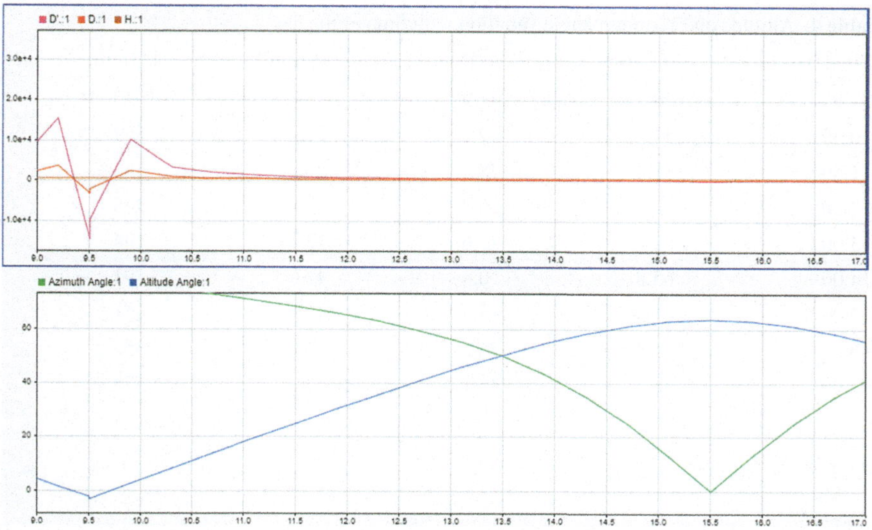

Fig. 18 Row Spacing Parameters for February

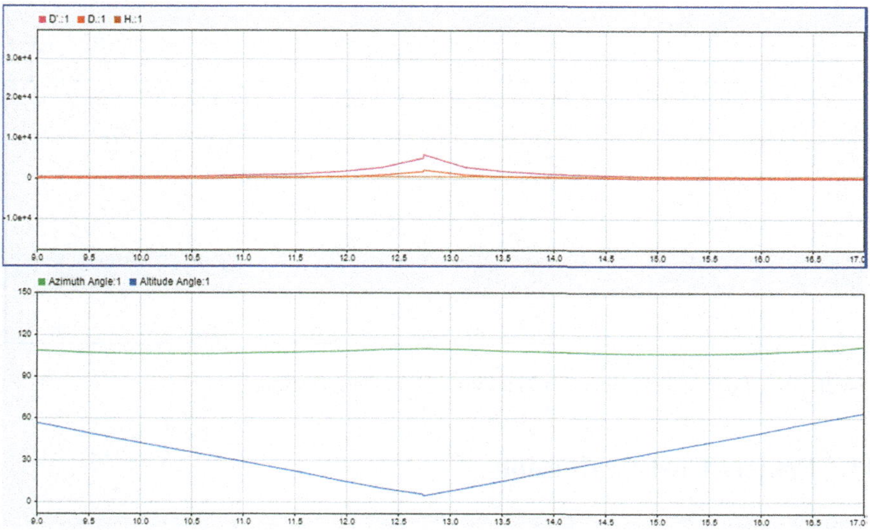

Fig. 19 Row Spacing Parameters for the July

vehicles handled by the plant every day can be calculated. Furthermore, the model provides the data for the variation in the Current, Voltage, Power generated by the modules at the plant throughout the year and at the same time the optimum row spacing with the land coverage required for setting up the charging station.

Table 4 Altitude and Azimuth angle variation with time of the day

Time (Hrs)	α (Deg.)	γ (Deg.)	H (mm)	D (mm)
09.00	29.7	47.9	868	1040
10.00	44.3	39.1	716	569
11.00	50.4	28	565	348
12.00	60.7	15.7	490	265
13.00	65.2	2.6	421	194
14.00	65.8	0.4	417	191
15.00	61.5	13.5	477	251
16.00	55.4	26	568	352
17.00	38.4	45.1	720	575

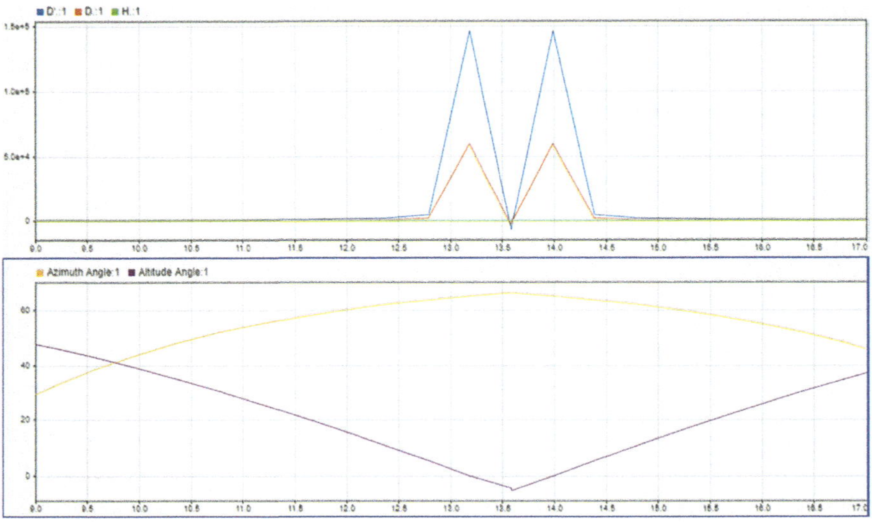

Fig. 20 Row Spacing Parameters on 21st December (Winter Solstice)

11 Project Cost Evaluation

11.1 Cost of Modules

Cost for every PV module = 9000 INR

The gross cost for 198130 module producing 11.25 Mega Watt = 1.8 billion INR

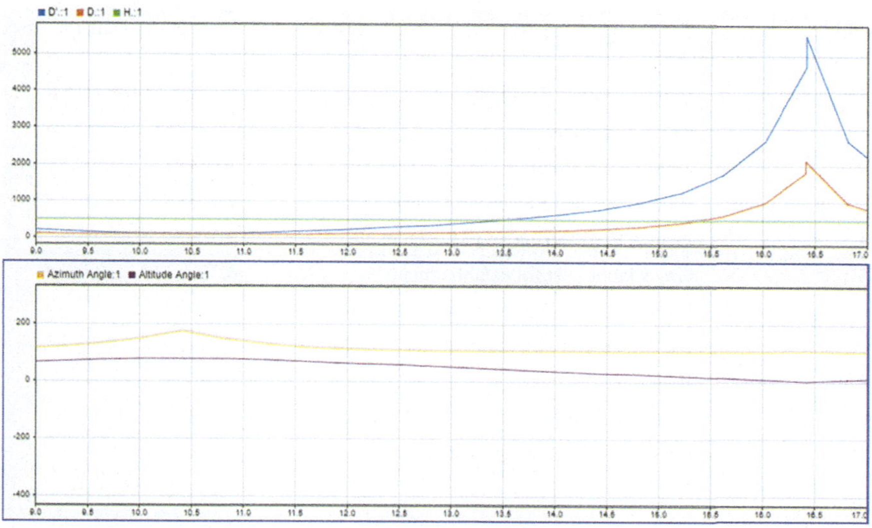

Fig. 21 Row Spacing Parameters on 21st June (Summer Solstice)

11.2 Design Engineering and Management Cost

Cost of man-power = Rs. 200−250/man-hour

Engineering-design and product-based planning = 2−3 h/day [7]

Gross cost for engineering-design and product-based planning for 11.25 MW = 4.5 million INR (Table 4).

11.3 Installation Workforce Cost

Wage for furnishing = Rs. 50–60/man-hr

Furnishing time required for every kWp = 12–13 man-hr

Gross manpower wages for the establishment of 11.25 MW power plant = 6.75 million INR (Tables 4 and 5).

11.4 Capacity Factor [21]

Capacity factor is considered as a significant factor to be determined when analyzing solar PV plant's money-making. Most of the expenditure for the power plant is usually settled, and the LCOE(i.e., Levelized Cost of Energy) would correspond to the usage of the energy generated at the power plant. Hence, the term Capacity Factor

Table 5 Gross cost for modules, batteries with other Equipment and miscellaneous charges used for Solar PV System [20]

Serial number	Items for Solar PV Plant	Price in Million (INR)
1	Cost of Modules	1800
2	Array Construction	104.445
3	Electronics and Electrical Accessories	145.35
4	Project Engineering and Management	4.5
5	Gross labor cost for establishment	6.75
6	Positioning Civil and Mechanical Equipment	22.5
7	Packaging and Shipment	2.25
8	Batteries	2500
9	Charge Controller	156.25
10	**Gross cost for charging station**	**4742.045**

is very suitable in the context of Solar Photovoltaic Plants for a glimpse of the actual energy production rate.

$$\text{Capacity Factor} = \frac{\text{Annual kilowatt} - \text{hours generated for each kilowatt peak Capacity}\left(\frac{\text{kWh}}{\text{kWp}}\right)}{\text{Hours Plant is used}}$$

The energy required to be generated = 30000 kW h per day
Annual energy requirement of the plant = 30000 × 365 = 10.95 * 10^6 kW-hr
Maximum capacity of the PV plant = 11.25 *10^3 kW_p
Therefore, CF is = 0.333 = 33.3%

12 Conclusion

Parking garages give an extraordinary area to sun oriented capacity to be placed into the electrical matrix on account of the gainful social estimations of shade and comfort for vehicle charging. For cases in which the separation to work requires charging to commute home electrically, there is a critical advantage related to charging while stopped at work. A considerable measure of intensity can be created utilizing sun-oriented boards in parking areas. The planning of intensity age with sun-powered fits well with the necessities of labourers with electric vehicles and 9:00 to 5:00 business. The Nissan Leaf with the sun based controlled charge station has the most minimal expense more than ten years, and the least ozone-depleting substance outflows contrasted with different choices that were researched. Its effect on urban air quality is likewise little. Since it has a constrained range, the chance to charge the vehicle at work has enormous worth. There are numerous reasons and advantages related to pushing ahead in the establishment of sun oriented fueled charging stations. The utilization of sunlight based fueled charge stations for electric vehicles has various benefits as far as natural, social, and monetary viewpoints. Sun based controlled charge stations, and electric vehicles at present do and will keep on profiting society in general.

Acknowledgements Authors sincere thanks to Prof. Raja Sekhar Y at Vellore Institute of Technology, Vellore for useful suggestions and comments, and allowing us to work at this project.

References

1. M.R. Rashel, A. Albino, T. Gonçalves, M. Tlemçani, Sensitivity analysis of environmental and internal parameters of a photovoltaic cell, in *Proceedings of the Energy for Sustainability International Conference Designing Cities & Communities for the Future Madeira* (2017)
2. H.H. Ozturk, D. Kaya,¨ *Electricity Generation from Solar Energy: Photovoltaic Technology* (Umuttepe Publications, Koca-eli, Turkey, 2013)
3. S. Patra, N. Kishor, S.R. Mohanty, P.K. Ray, *Power quality assessment in 3-U grid connected PV system with single and dual stage circuits* (2015)
4. A. Labouret, M. Villoz, *Energie photovoltaı¨que* (Dunode‘me e´dition Bernard Equer, Energie solaire photovoltaı¨que, 2006)
5. M. Šúri, T. Huld, E. Dunlop, PV-GIS: a web-based solar radiation database for the calculation of PV potential in Europe. Int. J. Sustain (2005)
6. R. Arulbel Benela, K. Jamuna. *Design of Charging Unit for Electric Vehicles Using Solar Power* (2018)
7. M. Chandel, G.D. Agrawal, S. Mathur, A. Mathur, Techno-economic analysis of solar photovoltaic power plant for garment zone of Jaipur city (2017)
8. S.M. Besarati, R.V. Padilla, D.Y. Goswami, E. Stefanakos, The potential of harnessing solar radiation in Iran: generating solar maps and viability study of PV power plants. Renew. Energy (2013)
9. S. Sánchez-Carbajal, P.M. *Rodrigo Optimum Array Spacing in Grid-Connect* (Photovoltaic Systems considering Technical and Economic Factors, 2019)
10. S. Juyal, H. Sanjeevi, A. Saxena, S. Sharma, A. Singh, Zero emission vehicles (zevs): towards a policy framework (2015)
11. N.D. Kaushika, N.K. Gautam, K. Kaushik, Simulation model for sizing of standalone solar PV system with interconnected array (2004)
12. H.A. Kazema, T. Khatiba, K. Sopianb, Sizing of a standalone photovoltaic/battery system at minimum cost for remote housing electrification in Sohar, Oman (2013)
13. D.A. Quansah, M.S. Adaramola, Comparative study of performance degradation in poly and mono-crystalline-Si solar PV modules deployed in different applications (2017)
14. G. Rainaa, S. Sinhab, Outlook on the Indian scenario of solar energy strategies: policies and challenges (2017)
15. S. Rauf, A. Wahab, M. Rizwan, S. Rasool, N. Khan, Application of dc-grid for efficient use of solar PV system in smart grid (2016)
16. A. Zakariazadeh, S. Jadid, P. Siano, Smart microgrid energy and reserve scheduling with demand response using stochastic optimization (2014)
17. G. Badea, R.A. Felseghi, M. Varlam, C. Filote, M. Culcer, M. Iliescu, M.S. Răboacă, Design and simulation of romanian solar energy charging station for electric vehicles (2018)
18. T. Huld, Software tools and data for the estimation of solar radiation and photovoltaic module performance over large geographical areas (2016)
19. H. Bellia, R. Youcef, M. Fatima, A detailed modeling of photovoltaic module using MATLAB (2014)

20. S. Ahsan, K. Javed, A.S. Rana, M. Zeeshan, Design and cost analysis of 1 kW photovoltaic system based on actual performance in Indian scenario (2016)
21. Capacity factor ⟨http://en.wikipedia.org/wiki/Capacity_factor⟩
22. S.P. Sukhatme, J.K. Nayak, Solar energy. McGraw-Hill Education (2017)
23. http://rolandvere.weebly.com/calculating-tilted-array-spacing.html. Accessed on 18-03-2020

Enhancement of IEEE 802.11 Based Network Performance Using Combined Optimization of Parameters for Smart Grid Network

Lipi Chhaya

1 Introduction

Wireless communication has become an indispensable and inevitable feature of various technologies. Protocols are structured using primitive layered methodology in which layers are intended for just passing the information to upper layers. With the advancement of data communication and real time applications, this approach has become insufficient for superior performance. For example, cognitive radio technology includes assignment of unoccupied channels pertaining to primary to secondary users for performance optimization. This technology necessitates real time observation of traffic parameters and channel conditions for bandwidth optimization [1–5].

Such kind of technology necessitates parameter optimization for performance enhancement. Joint optimization of parameters facilitates adaptive network performance. Moreover, cognitive radio technology is an unescapable constituent of Smart grid technology which integrates power and communication infrastructure for real time observation, control and management of energy parameters. Smart grid comprises of heterogeneous communication networks [6–10]. Smart grid communicatin infrastructure is an integration of several communication standards and technologies to serve numorous functionalities [11–14]. IEEE 802.11 is the most suitable standard for home area network of smart grid technology [6]. It facilitates real time monitoring and control of energy statistics [6]. This paper depicts simulation results of parameter optimization of IEEE 802.11 standard using PAMVOTIS software.

L. Chhaya (✉)
Researcher, Gandhinagar, Gujarat, India
e-mail: lipi.chhaya@gmail.com

V. K. Singh et al. (eds.), *Renewable Energy and Future Power Systems*, Energy Systems in Electrical Engineering, https://doi.org/10.1007/978-981-33-6753-1_8

2 Joint Optimization of IEEE 802.11 Parameters

This section illustrates the joint optimization of physical and media access layer parameters for performance improvement. Simulation is done using PAMVOTIS simulator to examine an effect of RTS/CTS mechanism as well as adaptation of EDCA parameters. IEEE 802.11 protocol is intended for WLAN. In the physical layer parameters, the modulation scheme, data rates etc. are optimized. IEEE 802.11 standard provides three options namely FHSS, DSSS and infrared. FHSS operates in ISM band using GFSK modulation techniques and provides the data rate of either 1 Mbps or 2 Mbps. DSSS functions in 2.4 GHz ISM band using DBPSK technique for 1 Mbps data throughput and DQPSK for 2 Mbps data throughput. Physical layer of WLAN standard is conceptually divided into two sub-layers namely PLCP and PMD sub-layer. PMD is associated with modulation and coding. PLCP is associated with the services which are offered to MAC layer by physical layer. PLCP handles Clear Channel Assessment (CCA) and service access point. CCA is used for CSMA/CA mechanism used by MAC layer [15].

Infrared alternative works using pulse position modulation scheme. In wired network such Ethernet, carrier sensing is quite simple as the peak voltage at the cable is compared with threshold value. Carrier sense is a complex task in wireless network. In IEEE 802.11, the PHY layer sensing is through CCA provided by PLCP. The assessment is performed either by measuring the received signal strength and comparing it with threshold value or by detection of bits. The process of received signal strength is faster but it is not as reliable as bit detection as the interference may cause false sensing.

The function of MAC layer is arbitration and multiplexing of various transmission requests of a specific area [16–18]. It also takes care of roaming, energy consumption and authentication. Distributed coordination function is based on CCA and CSMA/CA. It is inevitable for accessing IEEE 802.11 standard. CSMA-CA/CD is the well-practiced technique applied for wired networks. This technique cannot be used for wireless network as the error rate is much higher in WLAN. Detection of collision will drastically reduce the throughput in WLAN.

Thus, the technique of collision avoidance is applied for WLAN. EDCA mechanism is proposed for Quality of Service enhancements in WLAN in the standard IEEE 802.11e. This mechanism is an extension of basic Distributed Coordinated Function. It classifies the traffic into four access groups such as voice, video, best effort, and background. Higher priority is assigned to the traffic with higher QoS requirements. Inter Frame Spacing is the time span between two consecutive frames in WLAN. In WLAN, Arbitration Inter Frame Spacing (AIFS) is a methodology to prioritize one access category over other. It works by widening or shortening the time period during which a node has to wait before transmitting the next frame. A small value of AIFS means higher probability of the message to be transmitted.

2.1 RTS/CTS Parameter Optimization

The size of contention window must be carefully chosen as the too small size may result into packet collisions and larger size may result into unnecessary delay. Wireless networks are prone to the ''Hidden terminal'' problem. In this problem, one node receives information from two nodes which are unable to see each other. This situations gives rise to collisions and therefore reduced throughput. The transmitter nodes are unaware of this situation and in an impression that the communication process is successful.

RTS/CTS is used to assuage the hidden terminal issue. The transmitter transmits Request To Send packet to receiver which contains the address of the receiver of next data packets and the expected duration. The stations which receive this RTS, set a network allocation vector. The receiver sends Clear To Send message to transmitter to indicate that it is ready to receive the packet from transmitter. Once this RTS as well as CTS exchange process in completed, all the neighboring stations are informed that the medium is exclusively busy for one sender.

After the transmission process is over, the network allocation factor marks the channel as available. The chances of detectable collision are only possible when the RTS/CTS are being sent. Once the network allocation vector is set, collision will not exist. This process adds overhead bits. To determine the necessity of RTS/CTS, a specific value of RTS threshold should be selected. RTS/CTS is activated only if the frame size is more than threshold of RTS. The sequence will be RTS-CTS-DATA-ACK which is called as four way handshake method. If the frame size is less than the value of RTS threshold, then the two way handshake DATA-ACK is followed.

The effect of RTS/CTS threshold is analyzed for different RTS threshold values. Media access delay will increase as a result of RTS/CTS mechanism as the process of exchanging this commands will procure some time duration. The probability of using RTS/CTS mechanism is inversely proportional to the threshold value. The higher the value of RTS threshold, the lower the probability of sending RTS frames. The effect of RTS threshold is depicted using graphical representation. The representations show mean value of results. Total hundred values are derived per single node. It is apparent from the results that media access delay as well as total packet delay is the lowest in IEEE 802.11 standard without enabling this mechanism as the communication process is faster. For lowest RTS threshold, the probability of RTS/CTS is highest and therefore the delay is the highest. Total ten nodes are considered and hundred readings are taken for each node.

Table 1 shows the results for Mean Values of throughput for all 10 nodes. Figure 1 shows the Mean Values for different RTS thresholds. Results depicts that mean value of throughput is highest for lower value of RTS threshold as the lower value of RTS threshold ensures highest chances of sending RTS frame. Graphical results prove that the throughput and media access delay are higher in case of RTS threshold = 16. This is because the possibility of sending RTS frame is the maximum in this case compared to threshold values of 256 and 1024.

Table 1 Mean Values of results for throughput

Node	Mean values-basic-no RTS/CTS Throughput (Kbits/s)	Mean Values-RTS = 16 Throughput (Kbits/s)	Mean Values-RTS = 256 Throughput (Kbits/s)	Mean Values-RTS = 1024 Throughput (Kbits/s)
1	63	64	65	64
2	65	65	63	64
3	64	65	65	64
4	64	65	64	64
5	63	64	64	64
6	65	65	65	64
7	64	64	63	64
8	64	66	65	64
9	64	63	65	64
10	63	66	62	64
Average	63.9	64.6	64	64

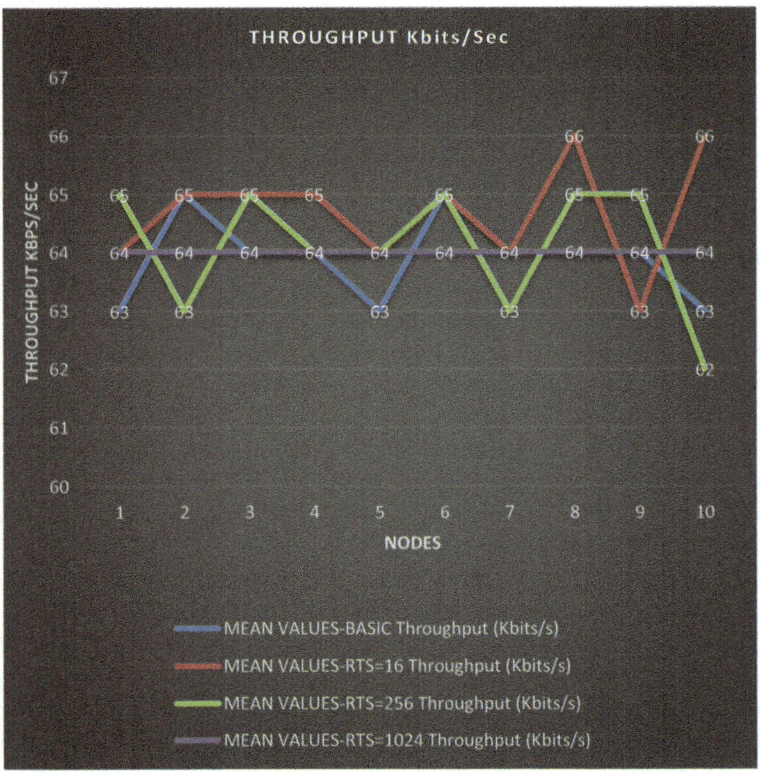

Fig. 1 Mean Values of results for throughput

Table 2 Mean Values of results for media access delay

Node	Mean Values-Basic	Mean Values-RTS = 16	Mean Values-RTS = 256	Mean Values RTS = 1024
	Media Access Delay (msec)	Media Access Delay (msec)	Media Access Delay (msec)	Media Access Delay (msec)
1	0.59	21.56	19.99	18.92
2	0.59	21.35	20.66	19.74
3	0.58	20.74	20.35	19.11
4	0.59	21.03	20.16	19.54
5	0.59	21.65	20.51	19.11
6	0.59	21.39	20.52	19.43
7	0.59	21.33	20.53	20.2
8	0.59	21.74	20.42	19.78
9	0.58	21.39	19.97	19.83
10	0.59	21.9	20.24	19.24

As shown in above graphs, media access delay is calculated for different nodes for various threshold values using simulations. Apart from distinct results, Mean Values are also calculated using simulations as illustrated in Table 2. Figure 2 shows the graphical results. It shows that media access delay is highest for RTS threshold value of 16.

The above results show that RTS mechanism increases media access delay. The highest delay is resulted for lowest value of RTS as it has the highest probability of sending RTS frame.

Table 3 shows the results of Mean Values of total packet delay. It is highest for RTS threshold value of 16. Figure 3 shows the graphical illustration of results.

The above results show that RTS mechanism increases total packet delay. The highest delay is resulted for lowest value of RTS as it has the highest probability of sending RTS frame.

2.2 EDCA Parameter Optimization

The standard as well as adapted values of AIFS are depicted in the table. Highest priority is assigned to the access category with smallest AIFS. Media access delay and total packet delays are also depicted using tabular and graphical representation. Table 4 shows the default values of EDCA parameters. Table 5 shows the diverse set of adapted values of EDCA parameters.

Table 6 shows the results for Mean Values of throughput for basic as well as adapted EDCA parameters. Throughput is maximum for adapted cases 2 and 5.

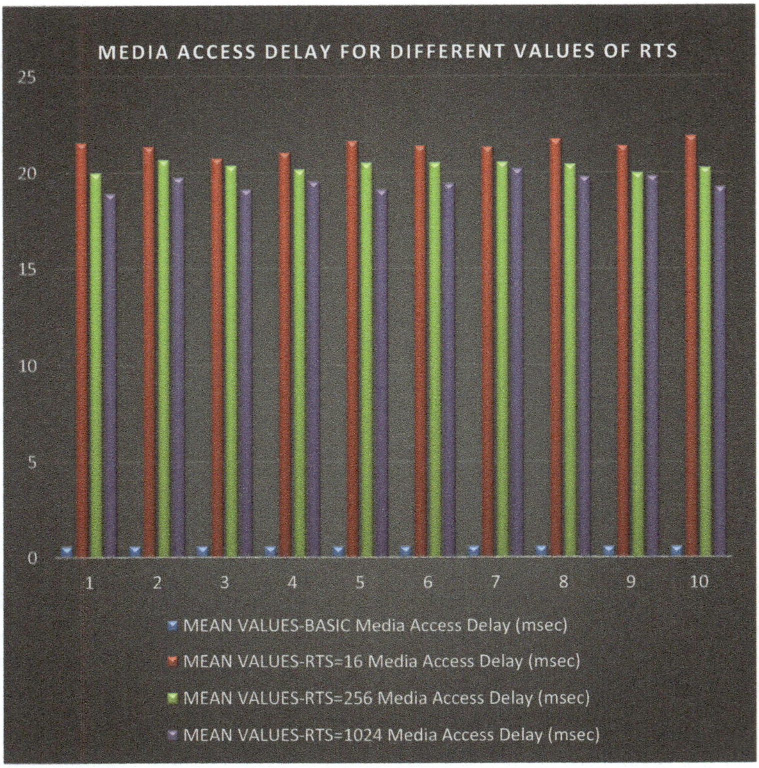

Fig. 2 Mean Values of results for media access delay

Table 3 Mean Values of results for Total Packet Delay

Node	Mean values Basic Total Packet Delay (msec)	Mean values-RTS = 16 Total Packet Delay (msec)	Mean values-RTS = 256 Total Packet Delay (msec)	Mean values-RTS = 1024 Total Packet Delay (msec)
1	0.59	31.56	28.64	24.71
2	0.59	29.41	28.33	26.08
3	0.58	28.13	28.32	26.16
4	0.59	29.69	28.04	25.77
5	0.59	29.93	29.73	26.06
6	0.59	30.88	28.46	25.79
7	0.59	30.11	29.86	27.09
8	0.59	30.36	29.26	27.21
9	0.59	30.54	26.94	26.94
10	0.59	32.26	27.53	26.42

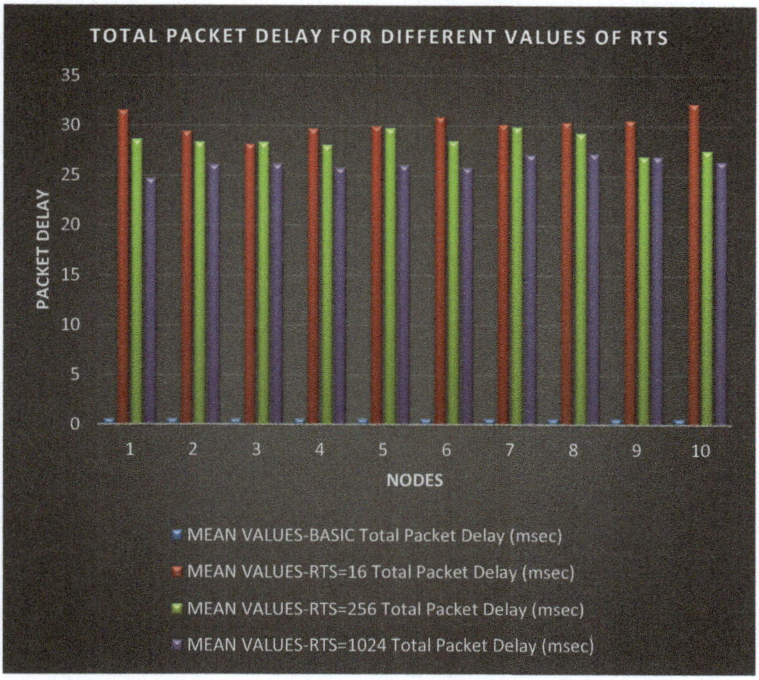

Fig. 3 Mean Values of result for Total Packet Delay

Table 4 Default values of EDCA parameters

Access category	Cwmin	Cwmax	AIFSN
AC_BK	15	1023	7
AC_BE	15	1023	3
AC_VI	7	15	2
AC_VO	3	7	2

Table 5 Default and adapted values of EDCA parameters for simulation

Type of EDCA	Basic-Default	Adapted EDCA-1	Adapted EDCA -2	Adapted EDCA -3	Adapted EDCA -4	Adapted EDCA -5	Adapted EDCA -6
AIFS (Ac0)	7	2	6	6	7	8	9
AIFS (Ac1)	3	2	3	6	7	6	6
AIFS (Ac2)	2	1	2	2	3	2	2
AIFS (Ac3)	2	1	1	1	2	1	1

Table 6 Mean Values of results

Node	Mean values-Basic-Throughput (Kbits/s)	Mean values-Adapted EDCA-1-Throughput (Kbits/s)	Mean values-Adapted EDCA-2-Throughput (Kbits/s)	Mean Values-Adapted EDCA-3-Throughput (Kbits/s)	Mean values-Adapted EDCA-4-Throughput (Kbits/s)	Mean values-Adapted EDCA-5-Throughput (Kbits/s)
1	63	65	65	63	63	65
2	62	63	64	65	63	64
3	65	64	64	64	65	64
4	65	63	65	63	65	65
5	64	64	63	64	63	63
6	64	64	65	65	65	65
7	64	64	64	64	65	64
8	64	65	64	64	64	64
9	64	65	65	64	64	65
10	64	64	63	64	64	63
Average	63.9	64.1	64.2	64	64.1	64.2

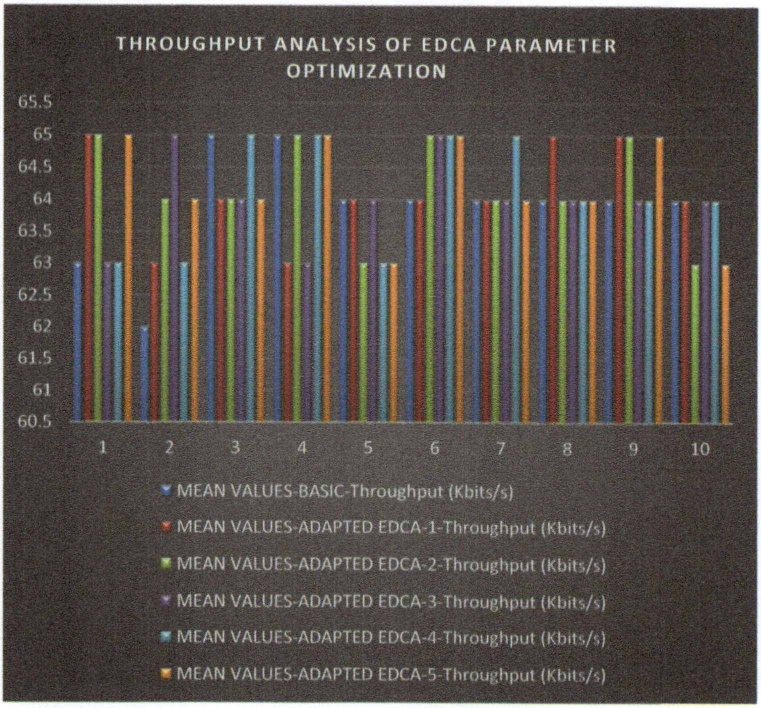

Fig. 4 Mean Values of results for throughput

Figure 4 illustrates that adaptation of EDCA parameters results into enhanced throughput. The highest throughput is achieved in adapted EDCA-2 and 5. It is the lowest in case of default values. The following graphs show the results of media access delay.

Table 7 shows the Mean Values of media access delay for default and adapted EDCA parameters. As shown in tabular results, it is the lowest for adapted EDCA-3.

Results illustrated in Fig. 5 show that the lowermost value of MAC delay is achieved for adapted EDCA-3. The next section shows the graphical results of media access delay for various EDCA parameters.

Table 8 shows the Mean Values of total packet delay for basic as well as adapted EDCA parameters. Tabular results illustrate that total packet delay is lowest for adapted EDCA-3 and highest for adapted EDCA-1.

Figure 6 shows the graphical results for Mean Values of total packet delay for different EDCA parameters. Results show that the lowermost value of MAC delay is achieved for adapted EDCA-3. From above outcome, it is apparent that a specific value of optimized parameter should be selected by considering varied set of results.

Table 7 Mean Values of Media access delay

Node	Mean values-Basic-Media Access Delay (msec)	Mean values-Adapted EDCA-1-Media Access Delay	Mean values-Adapted EDCA-2-Media Access Delay (msec)	Mean values-Adapted EDCA-3-Media Access Delay (msec)	Mean values-Adapted EDCA-4-Media Access Delay (msec)	Mean values-Adapted EDCA-5-Media Access Delay (msec)
1	4.79	5.09	4.76	4.55	5.11	4.76
2	4.78	5	4.83	4.54	4.91	4.83
3	4.79	4.98	4.84	4.5	4.89	4.84
4	4.89	4.97	4.82	4.55	4.88	4.82
5	4.78	4.94	4.81	4.5	4.94	4.81
6	4.85	4.95	4.84	4.63	4.91	4.84
7	4.79	4.94	4.87	4.52	4.97	4.87
8	4.73	4.93	4.86	4.61	4.89	4.86
9	4.91	4.93	4.89	4.51	4.82	4.89
10	4.87	5.02	4.88	4.48	4.92	4.88

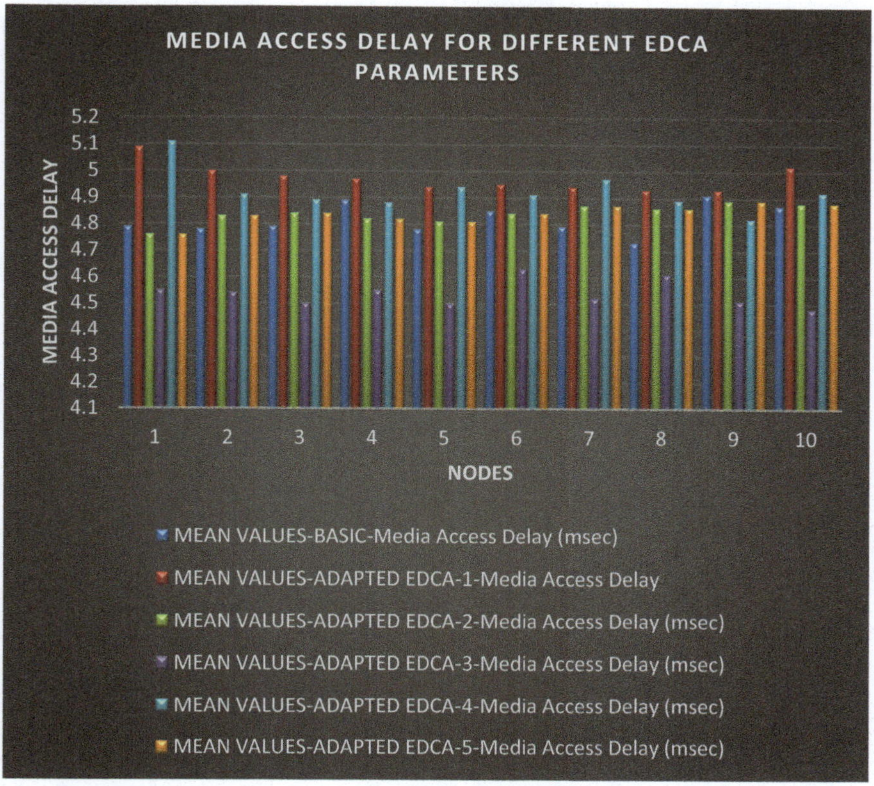

Fig. 5 Mean Values of Media access delay

3 Results and Discussion

A cross layer methodology is a defilement of outmoded layered design methodology through "Collaborative Optimization". It is apparent from the results that joint optimization of PHY and MAC layer parameters significantly enhances network performance. The above results show that RTS mechanism increases total packet delay. The highest delay is resulted for lowest value of RTS as it has the highest probability of sending RTS frame. Results illustrates that adaptation of EDCA parameters results into enhanced throughput. The highest throughput is achieved in adapted EDCA-2 and 5. It is the lowest in case of default values.

Table 8 Mean Values of total packet delay

Node	Mean values-Basic-Total Packet Delay (msec)	Mean values-Adapted EDCA-1-Total Packet Delay	Mean values-Adapted EDCA-2-Total Packet Delay (msec)	MEAN values-Adapted EDCA-3-Total Packet Delay (msec)	Mean values-Adapted EDCA-4-Total Packet Delay (msec)	Mean values-Adapted EDCA-5-Total Packet Delay (msec)
1	5	5.32	4.99	4.69	5.33	4.99
2	4.96	5.21	5.03	4.73	5.12	5.03
3	4.99	5.2	5.01	4.66	5.14	5.01
4	5.1	5.15	5.02	4.78	5.08	5.02
5	4.97	5.16	5.01	4.63	5.18	5.01
6	5.03	5.16	5.09	4.82	5.16	5.09
7	5.01	5.19	5.07	4.72	5.22	5.07
8	4.95	5.19	5.08	4.83	5.1	5.08
9	5.13	5.17	5.1	4.65	5.03	5.1
10	5.05	5.27	5.06	4.63	5.12	5.06

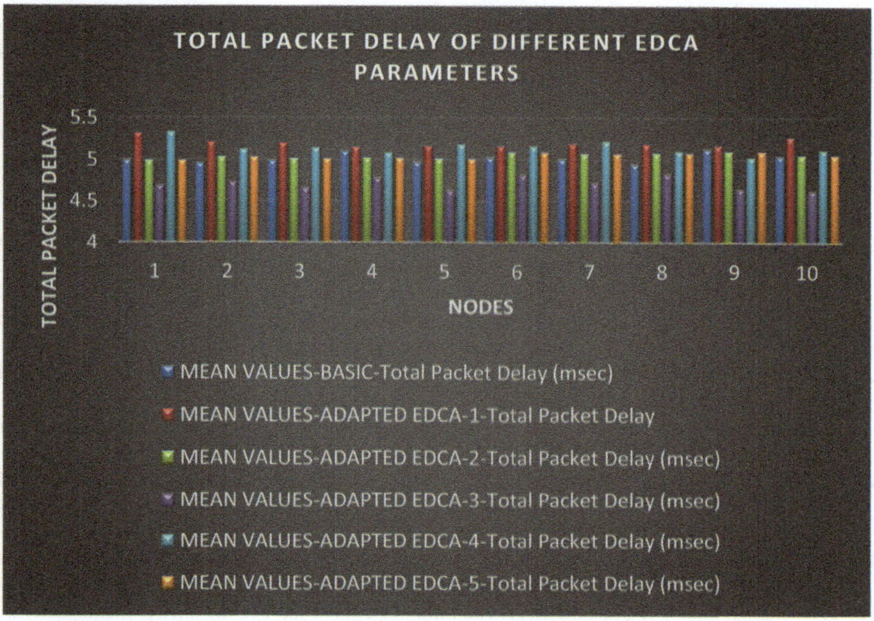

Fig. 6 Mean Values of total packet delay

4 Conclusion

This paper depicts joint optimization of PHY and MAC parameters of IEEE 802.11 standard. Diverse set of results are attained for different parameters like total throughput, MAC delay and total packet delay. Graphical results illustrate that the throughput and media access delay are higher in case of RTS threshold of 16. Tabular and graphical results illustrate that total packet delay is lowest for adapted EDCA-3 and highest for adapted EDCA-1. Results illustrate that the lowermost value of MAC delay is achieved for adapted EDCA-3. It is apparent from the results that joint optimization augments the performance of a wireless network. This paper is anticipated to serve as a directive for joint optimization of various parameters for different standards as well as applications for future research endeavors.

References

1. M.I.N. Binti, T.C. Wei, A.H.M. Yatim, Smart grid technology: Communications, power electronics and control system, in *International Conference on Sustainable Energy Engineering and Application (ICSEEA)* (Bandung, 2015), pp. 10–14
2. L. Chhaya, P. Sharma, A. Kumar, G. Bhagwatikar, Integration of Cognitive Radio with Heterogeneous Smart Grid Communication Architecture, in *Intelligent Communication, Control and Devices*. Advances in Intelligent Systems and Computing, vol. 624 (Springer, Singapore, 2018)

3. M. Erol-Kantarci, H.T. Mouftah, Energy-efficient information and communication infrastructures in the smart grid: a survey on interactions and open issues. IEEE Comm. Surveys Tutorials **17**, 179–197 (2015)
4. V.C. Gungor, D. Sahin, T. Kocak, S. Ergut, C. Buccela, C. Cecati, G.P. Hancke, Smart grid technologies: communication technologies and standards. IEEE Trans. Ind. Inform. **7**, 529–539 (2011)
5. Y. Yan, Y. Qian, H. Sharif, D. Tipper, A survey on smart grid communication infrastructures: motivations, requirements and challenges. IEEE Comm. Surveys Tutorials **15**, 5–20 (2013)
6. L. Chhaya, P. Sharma, G. Bhagwatikar, A. Kumar, Development of wireless data acquisition and control system for smart microgrid, in *Advances in Smart Grid and Renewable Energy*. Lecture Notes in Electrical Engineering, vol. 435 (Springer, Singapore, 2018)
7. S. Bera, S. Misra, M.S. Obaidat, Energy-efficient smart metering for green smart grid communication, in *IEEE Global Communications Conference* (Austin, 2014) pp. 2466–2471
8. J. Momoh, *Pathway for Designing Smart Grid*. Smart Grid: Fundamentals of Design and Analysis (IEEE Press, Wiley, 2012), pp. 122–139
9. L. Chhaya, P. Sharma, G. Bhagwatikar, A. Kumar, Communication theories and protocols for smart grid hierarchical networks. J. Elecr. Electr. Eng. 43–48 (2017)
10. L. Chhaya, P. Sharma, G. Bhagwatikar, A. Kumar, Wireless sensor network based smart grid communications: cyber-attacks, intrusion detection system, and topology control. Electronics **6**(1), 1–22 (2017)
11. I.S. Bayram, I. Papapanagiotou, A survey on communication technologies and requirements for internet of electric vehicles. J. Wireless Commun. Netw. **2014**, 223 (2014)
12. J. Huang, H. Wang, Y. Qian, Smart grid communications in challenging environments, in *IEEE Third International Conference on Smart Grid Communications* (Tainan, 2012), pp. 552–557
13. L. Chhaya, P. Sharma, A. Kumar, G. Bhagwatikar, *Cybersecurity for Smart Grid: Threats, Solutions and Standardization*. Advances in Greener Energy Technologies (Springer, Singapore, 2020), pp. 17–29
14. L. Chhaya, P. Sharma, A. Kumar, G. Bhagwatikar, *EMI Concerns, Measurements and Standards for Smart Grid*. Advances in Greener Energy Technologies (Springer, Singapore, 2020), pp. 123–136
15. IEEE Wireless LAN Medium Access Control (MAC) and Physical Layer (PHY) specifications, IEEE 802.11 standards (1999)
16. IEEE Std 802.1 XTM-2004, IEEE Standard for Local and Metropolitan Area Networks:Port-Based Network Access Control (2004)
17. IEEE Std 802.11TM-2007, IEEE Standard for Information Technology–Telecommunications and Information Exchange Between Systems–Local and Metropolitan Networks-Specific Requirements. Part 11: Wireless LAN Medium Access Control (MAC) and Physical Layer (PHY) Specifications (2007)
18. C.U. Wang, H.Y. Wei, IEEE 802.11n MAC enhancement and performance evaluation. Monet **14**,760–771 (2009)

Wind Turbines in Energy Conversion System: Types & Techniques

Bibhu Prasad Ganthia, Subrat Kumar Barik, and Byamakesh Nayak

1 Introduction

The demand for global energy led to a reduction in fossil fuel supply and an increase in CO_2 levels. A great many countries have changed their energy policies to include more efficiency, due to the harmful effects of the carbon emissions on the environment. Because of the ubiquity, and because wind energy is an choice for many countries to contribute and to increase the dependence on an electric power network. The 22,446 MW installed offshore wind power generation in 2014 accounted for a fraction of the world's total installed wind power capacity. In 2050, the wind turbine could generate 25-30 percent of the global grid. The wind power production capacity installed in India is 37693.75 MW and is expected to rise to 60,000 MW by 2022 [1, 2].

2 Wind Energy Conversion System

Wind is created by irregular atmospheric heating of sunlight. The creation of hot and cold air regions results in a pressure gradient. The air flow from a region of high pressure to the region of low pressure cause the wind flow. Energy in motion is called the kinetic energy of water. The wind power conversion network block diagram with all parts is shown in Fig. 1. The aerodynamically engineered blades, which results

B. P. Ganthia (✉) · S. K. Barik · B. Nayak
KIIT Deemed to Be University, Bhubaneswar, Odisha, India
e-mail: jb.bibhu@gmail.com

S. K. Barik
e-mail: skbarikfel@kiit.ac.in

B. Nayak
e-mail: bnayakfel@kiit.ac.in

V. K. Singh et al. (eds.), *Renewable Energy and Future Power Systems*, Energy Systems in Electrical Engineering, https://doi.org/10.1007/978-981-33-6753-1_9

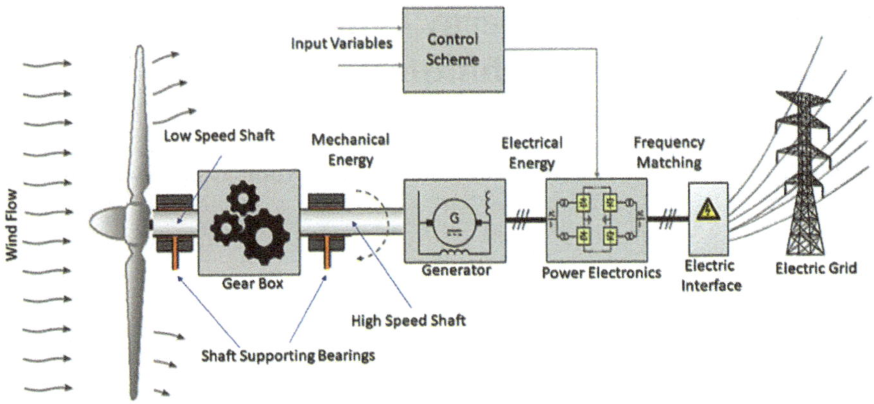

Fig. 1 Wind energy conversion system

in the wind turbine rotation, are influenced by this cinematic energy. Often the wind velocity is not enough to move the blades at a rate that cannot produce electricity. In some of WECS topologies, gearbox is helps to raise the speed of the rotor shaft. The induction generators used to convert mechanical energy of the shaft into electrical energy. This electrical energy is delivered to the grid [1].

Every wind turbine is a mechanical motor that includes aerodynamic blades to extract and transform power from a turbulent motion, the wind. The available power will rely upon the wind speed so it is important that the power at different wind speeds is controlled and restricted in order to avoid grid disruption. In general they are the three most common forms of conventional wind generator systems. One type is a continuous wind turbine system with a regular squirrel cage induction generator (SCIG), which is connected to the power supply network explicitly. The second type is a wind turbine variable speed system with Double Fed Induction Generator (DFIG) [2, 3]. The power rate of the winding rotor electric converter is approximately 30%; the DFIG winding stator is connected directly to the grid. The third type is a variable wind turbo with a fully-rated, synchronous generator or SCIG electronic conversion system. Usually, two generators use a multi-stage transmission. Synchronous generators can be driven directly by low ratio gear box systems, such as permanent synchronous magnet generator (PMSG). An interesting alternative is one or two-phase gear box [4].

3 Wind Turbine System

The power level for the electrical converter supplying the winding rotor is around 30 percent; the winding statistics for the DFIG are connected directly to the grid. The third category is a variable speed wind turbine with a full-speed electronic transmission system and a synchronous generator or SCIG. Usually, the two first

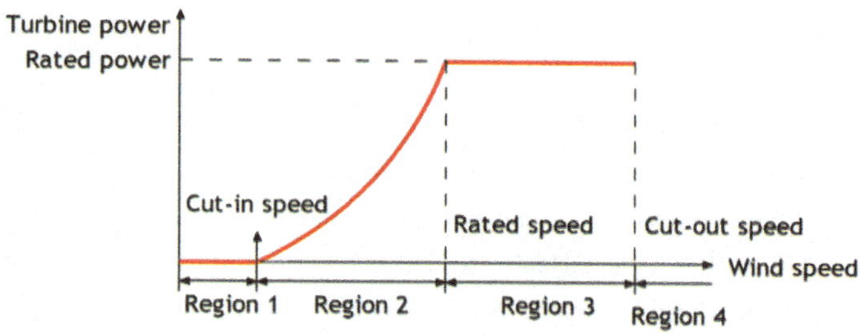

Fig. 2 Power curve for wind turbine system

generators use a multi-stage gearbox. Synchro generators, like a PMSG, can be powered directly by the low-ratio gearbox system; an interesting feature is one or two gearboxes. With the decline in the cost of electronic power converters, the B2B wind turbine is becoming a great option. Direct PMSG-based configurations are an attractive alternative for WECS compared to other wind turbine generators with a variable rpm. In contrast to the DFIG wind turbine system, the WT-PMSG does not necessitate a gearbox and slip-ring to allow high-performance WECS construction, stable operation and low maintenance. In the wind turbines industry, the efficient generation of more power by onshore and offshore wind turbines using PMSGs is continuously being investigated [5, 6].

Figure 2 illustrates the power curve into four areas. Region I: Standby, Region II: full power regulation, Region III: regulation of power rating and Region 4: shut-down of turbines. Region I: Standby type. Considering that Region II has low wind speeds and its blades below turbine capacity, maximum power can be extracted at the highest output. Area 3 is a control region that mainly conserves low noise and rotor torque. The pitch angle in Region III is isolated from its maximum power conversion angle by power measurement [7, 8] (Table 1).

The model is revised to reflect changes with the evolution of the technology. Each sort of generic WTG has one to four models, depending on the topology. These models are linked together for the simulation of events in the standard transient stability study, such as 20 to 30 s. The table no. 1 summarizes the popular WTG versions which are now available for industrial applications at onshore and offshore areas.

4 Type I Wind Turbine

Type-I wind turbine system shown in Fig. 3 is basically common and based on fixed or constant speed operated. This can be designed in stand alone or grid connected mode using a step up high rated power transformer. In this wind turbine system an squirrel cage induction generator (SCIG) which is operated in asynchronous mode is directly connected with the grid with constant speed controlled devices. This type of

Table 1 Models of wind turbine system available

Model	Description
Generic Wind Generator Model	
WT1G1	Type I generator system direct connected
WT2G1	Type- II Induction generator based external rotor resistor
WT3G1	Type-III WTs with Doubly-fed induction generator
WT3G2U	Type-III and Version II, Doubly-fed induction generator
WT4G1	Type-IV and Version II, Type-IV WTs with power converter
W4G2U	Power Converter built in Wind Turbine Systems
Generic Wind Electrical Model	
WT2E1	Type-II Wind Turbine with Rotor resistance control
WT3E1	Type III wind turbine with electrical controllers
WT4E1	Type IV wind turbine with electrical controllers
W4E2U	Type IV version II wind turbine with electrical controllers
Generic Wind Mechanical Model	
WT12T1	Type I and Type II WTs with 2-Mass Model
WT3T1	Type III WTs with Mechanical Model
Generic Wind Pitch Control Model	
WT3P1	Type III Wind Turbine with Pitch control model
Generic Wind Aerodynamic Model	
WT12A1	Type I and Type II with Pseudo-governor model
Voltage and Frequency Protection Model	
FRQDCA	Under/Over frequency generator with normal relay WT system
FRQTPA	Under/Over frequency generator with trip relay WT system
VGTDCA	Class I WTs with Under/Over frequency generator with disconnection relay
VTGTPA	Class II WTs with Under/Over frequency generator with trip relay

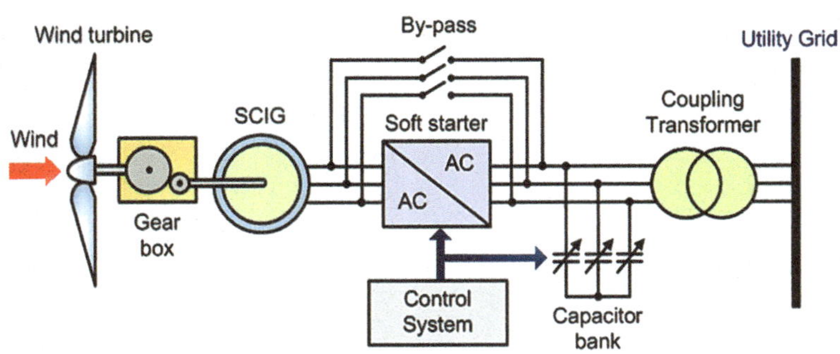

Fig. 3 Type I wind turbine system (squirrel cage induction generator with fixed speed operation)

induction generator draws reactive power when it is connected to the AC grid system with a capacitor used as condenser which helps in minimizing the reactive power effect in the grid system. This condenser of capacitor bank switched continuously for the reactive power compensation and act as a soft starter.

Fixed speed or constant speed operated wind turbine system mechanisms have many useful activities as it is very lost cost, availability in the market is nominal and maintenance free. The disadvantage of Type-I wind turbine system is it is unable to withstand for long duration with affixed speed of rotations. This instability due to the speed control makes the grid unstable. This method cannot maintain the constant speed for the grid system. So the Type-II wind turbine system coming into the feature with variable speed and constant voltage and frequency for withstanding the high mechanical strength and stress [9, 10].

5 Type II Wind Turbine System

This Type-II wind turbine system coming into feature due to no control over the speed in Type-I wind turbine system. In this type of turbine system wound rotor induction generators (WRIG) are used partial variable wind speed rotating system is present with use of variable rotor resistance. In the Fig. 4 shown below the model is used type-II turbine system with grid integrations is presented. Here in this model a rotor resistance in connected for the rotor speed control during wind speed variations and can control the speed of the rotor. A capacitor bank as condenser helps in reactive power compensation and acts as a soft starter for the stability. The converter control system is implemented in the model for the rotor side converter control on speed [10].

This featured win turbine system uses the converter control mechanism with the help of rotor resistance to control the rotor speed during grid integrations. This means that the rotor resistance can be controlled, but that the optically patented system

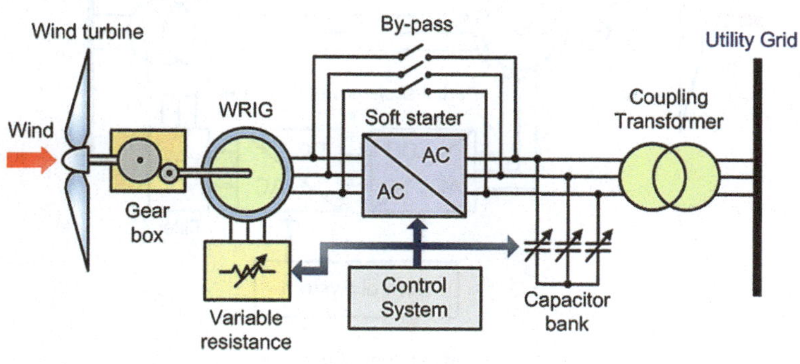

Fig. 4 Type II wind turbine system (wound rotor induction generator operated by variable rotor resistance with Partial variable speed)

prevents the need for costly ring-slipping that requires breasts and maintenance. The slip and thus the wind turbine power output can be operated by varying the rotor resistance. With the size of the variable rotor resistance, the dynamic speed control range is usually up to 10 percent higher than the synchronic speed. The heat loss energy from the external power converter is dumped. The disadvantage of this model is partial control on rotor speed hence the Type-III wind turbine system is coming into features [8, 10].

6 Type III Wind Turbine System

The DFIG rule is based on a specific wind turbine operated by speed and a WRIG and a partial power converter (rated approximately 30 percent of nominal value) on the rotor circuit shown in Fig. 5. Using electronic electricity converters, wind turbines run at variable (or adjustable) speeds, allowing more effective power recovery than fixed speed system. Other important benefits include the reduction of mechanical losses which make for lighter mechanical structures and a more active power output; cost is very efficient, smooth pitch control, power quality enhancement and faster performance, reduced high noise level and can be operated in islanding mode.

As shown in the Fig. 6 the rotor control and stator control is linked to the grid directly and the rotor frequency and thus the rotor speed are controlled through a partial power converter. In addition, this converter helps in the compensation of reactive power into the grid and the power flow in the grid fully controlled by Type-III wind turbine system. This method is economically desirable because of the smaller power converter with faster response. The main disadvantage of Type-III wind turbine

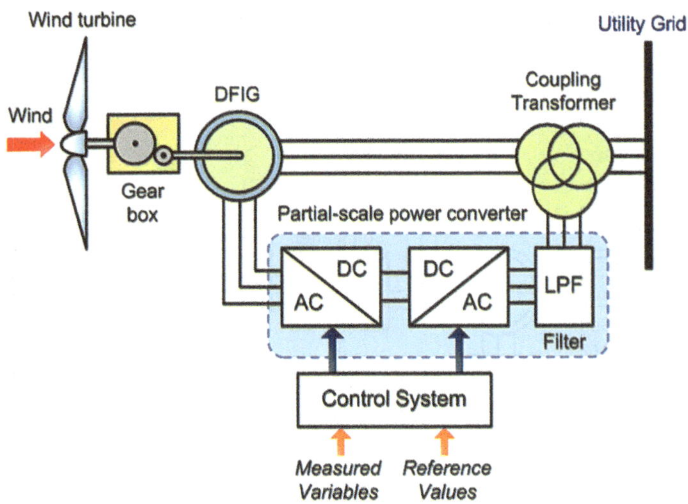

Fig. 5 Type III wind turbine system (DFIG operated by partial-scale power converter)

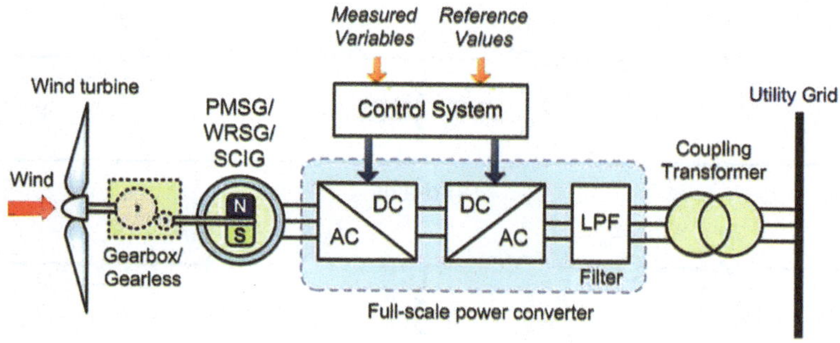

Fig. 6 Type IV wind turbine system (PMSG and full-scale power converter)

system is absence of brushes and it uses slip rings which need heavy maintenance and this system needs high protective systems. With this drawbacks of DFIG based wind turbine system Type-IV wind turbine system is coming into features. But the Type-III DFIG based wind turbine system effective and used for both onshore and offshore applications. The grid protection system also taken care in Type-III wind turbine system using various control techniques [9, 10].

7 Type IV Wind Turbine System

This type of wind turbine system is advanced and is fully controlled over rotor side and grid side. In this model permanent magnet synchronous generator is used for reducing maintenance like in Type-III slip rings. This model also can be operated in standalone as well as grid integrated mode. The advantages of this model are PMSG has brushes which demand less maintenance. The rotor side controller can control the pitch angle and the grid side converter controller can helps in controlling the reactive power using the adaptive techniques. The direct on line system helps to run the system in both synchronous and asynchronous modes. The generator system can excite by PMSG or WRSG which develop maximum efficiency in the wind turbine system during grid integrations.

The model is shown in the Fig. 6 where the all controllers are inserted in the system for faster response and higher control over variable speeds. This wind turbine system needs larger permanent magnets which bare very much costly and for the full converter control this demands faster protective system with hybrid systems that can control as well as maintain the safety measures for weak grid system [10, 11] (Fig. 7).

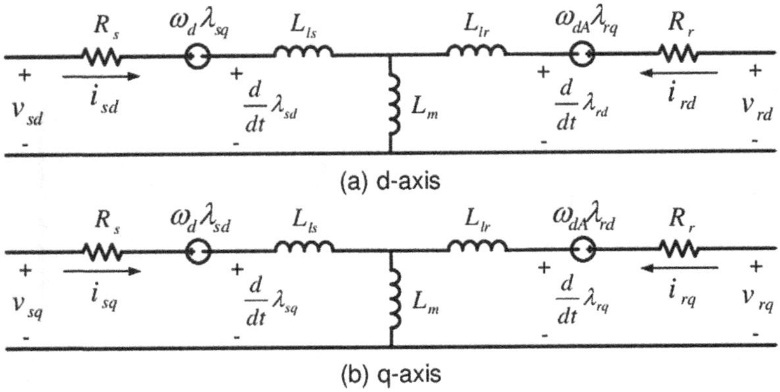

Fig. 7 Equivalent circuit of DFIG in: **a** d-axis and **b** q-axis

8 DFIG Modeling

The model in Fig. 8 illustrates a DFIG phasor diagram based on GSC (grid side control technique) functions of the voltage control (stator) Mechanism on the RSC (rotor side control). This control uses a PI control conventional system for the steady state analysis and transient analysis with settling times for the removal of faults during the grid integrated sudden wind speed variations [1, 12].

$$v_{ds} = -R_s x i_{ds} + (x_s + x_m) x i_{qs} + x_m x i_{qr} \tag{1}$$

$$v_{qs} = -R_s x i_{qs} - (x_s + x_m) x i_{ds} + x_m x i_{dr} \tag{2}$$

$$v_{dr} = -R_r x i_{dr} + (1 - \omega) x((x_r + x_m) x i_{qr} + x_m x i_{qs} \tag{3}$$

$$v_{qr} = -R_r x i_{qr} + (1 - \omega) x((x_r + x_m) x i_{dr} + x_m x i_{ds} \tag{4}$$

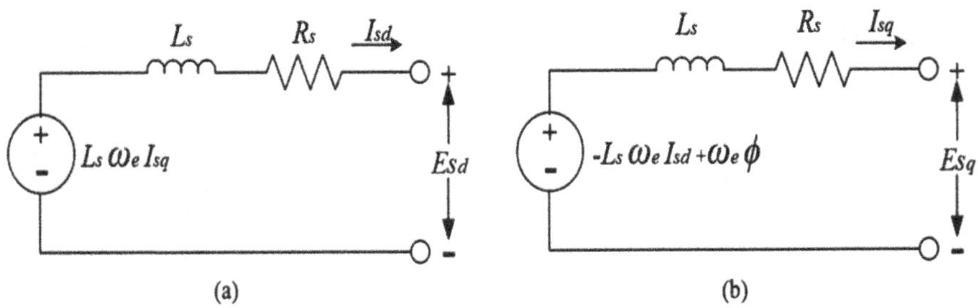

Fig. 8 d axis and q axis phasor diagram of PMSG

where: υ_{ds}, υ_{qs} are d and q axes of the stator voltages; υ_{dr}, υ_{qr}, d and q axes of the rotor voltages; ids, iqs, d and q axes of the stator currents; i_{dr}, i_{qr}, d and q axes of the rotor currents; R_a, R_r, stator and rotor resistance; xs, stator self-reaction; x_r, rotor self-reaction; x_m, rotor speed. The inserted active and reactive forces in the network can be expressed as given blow:

$$P = v_{ds} x i_{ds} + v_{qs} x i_{qs} + v_{dr} x i_{dr} + v_{qr} x i_{qr} \tag{5}$$

$$Q = -\frac{x_m v i_{dr}}{x_s + x_m} - \frac{v^2}{x_m} \tag{6}$$

Here is the magnitude of grid voltage.

$$\left.\begin{aligned} J\frac{dw_m}{dt} = T_e - T_m \\ T_e = \frac{3P}{2}\mathrm{Re}(j\vec{\lambda}_s\vec{i}_s^{\,*}) = -\frac{3P}{2}\mathrm{Re}(j\vec{\lambda}_s\vec{i}_r^{\,*}) \end{aligned}\right\} \tag{7}$$

Here J = MI-moment of inertial of the rotor (Kgm^2)

P = number of poles

Tm = Mechanical Torque (N.m)

Te = Electromagnetic Torque(N.m)

w_m = Rotor Mechanical Speed (rad/s).

The wind turbine system mechanical torque T_m is:

$$T_m = \frac{p_w}{\omega} \tag{8}$$

The wind turbine system power as P_w is:

$$p_w = \frac{1}{2}\rho c_p A * V_w^3 \tag{9}$$

where: c_p—power coefficient, ω—rotor speed, ρ—air density (kg/m^2), A—rotor area (m^2) and V_w (m/s)—wind velocity. Thus the electrical torque T_e and the mathematical link between T_m and T_e results:

$$T_e = x_m(i_{qr}i_{ds} - i_{dr}i_{qs}) \tag{10}$$

$$\frac{d\omega}{dt} = \frac{1}{2H}(T_m - T_e) \tag{11}$$

Here H is denotes rotor inertia [1, 13].

9 PMSG Modeling

A standard circuit diagram with an armature function along with the back electromotive forces (emf) can be used to electrically characterize the permanent magnet synchronously with normal condition. This model predicts equilibrium is deficient, the emf generated are sinusoidal so hysteresis and eddy current losses can be neglected with current dynamics. Hence the voltage equations are:

$$\begin{bmatrix} Uam \\ Ubm \\ Ucm \end{bmatrix} - \begin{bmatrix} Ua \\ Ub \\ Uc \end{bmatrix} = (Rm + sL)\begin{bmatrix} iam \\ ibm \\ icm \end{bmatrix} \tag{12}$$

$$Rm = \begin{bmatrix} Rm & 0 & 0 \\ 0 & Rm & 0 \\ 0 & 0 & Rm \end{bmatrix} L = \begin{bmatrix} Laa\ Lab\ Lac \\ Lab\ Lbb\ Lbc \\ Lac\ Lbc\ Lcc \end{bmatrix} \tag{13}$$

Here s is the laplace transform for $t > 0$, U is the voltage notation and R is the resistance. L is the mutual reactance with all combinations of the matrix representations.

PMSG uses permanent magnets which are helps in developing fluxes and it converts aerodynamic operated mechanical power to the electrical power in the system and transfer the power to the grid during grid integrations. This system uses converter models at both the rotor and grid side for the faster switching with a dc link as filter. The equivalent circuit diagram presented above in the figure no. 8. The voltage at the stator with respect to the rotating frame shown below as:

$$E_{sd} = -R_s I_{sd} - L_s(d/dt\,(I_{sd})) + \omega L_s I_{sq} \tag{14}$$

$$E_{sq} = -R_s I_{sq} - L_s\left(d/dt\left(I_{sq}\right)\right) - \omega L_s I_{sd} + \omega_e \varphi \tag{15}$$

where E_{sd}, E_{sq}, are d and q axes voltages, I_{sd}, I_{sq} are currents. R_s is resistance, L_s is self-inductance, φ is flux linkage and ω_e is rotational speed of PMSG.

The PMSG electromagnetic torque is:

$$T_e = 3/2\left\{pI_{sd}((L_d - L_q)I_{sd} + \phi)\right\} \tag{16}$$

where L_d, L_q are inductance, p is pole pairs [14–16].

10 Simulink Results

A constant wind speed (13 m/ s) was used to run the wind turbine. The valued value was selected for this wind speed. Voltage sags were simulated on the grid and the wind turbine's true and reactive power response observed. Net tension is reduced by 1 p.u. 0.8 p.u. 0.8 p. The sag lasts 18 cycles (0.3 s) at t = 15 s. The test aims to show that the model is effectively responsive to events in the dynamic time frame and that the machine's response to this event is realistic. The changes in grid voltage when the sag begins and ends are expected to result in an immediate answer. Note that as predicted from a fixed-speed wind turbine, the velocity does not change by much (roughly 2%). The actual power and reactive power outputs are disturbed, and the outputs show that after the ends of the sag, a mechanical oscillation happens and eventually the oscillation dampens [17, 18].

10.1 Type-I FSIG WTs Simulink Model and Results

The wind turbine model with mathematical parameter calculation block diagram represented in the Fig. 9.

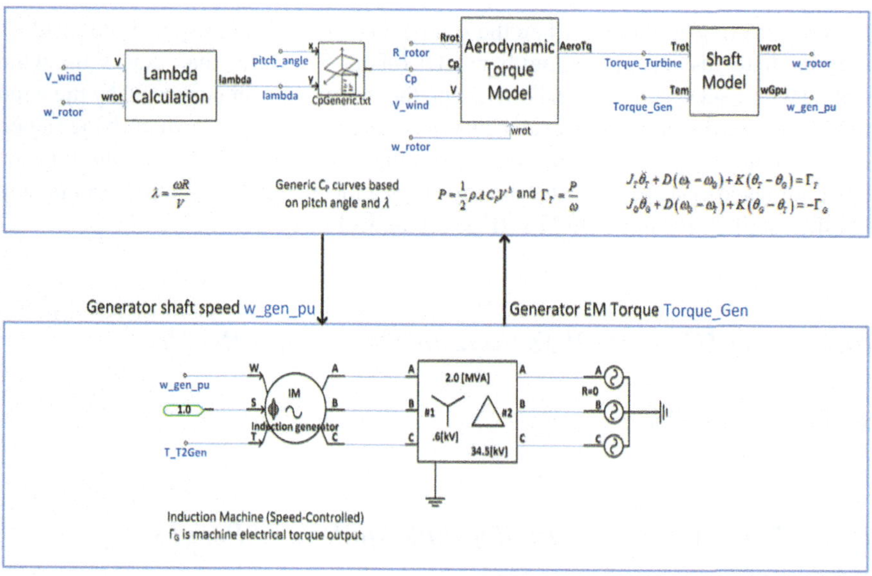

Fig. 9 Mathematical Modeling of Wind Turbine system

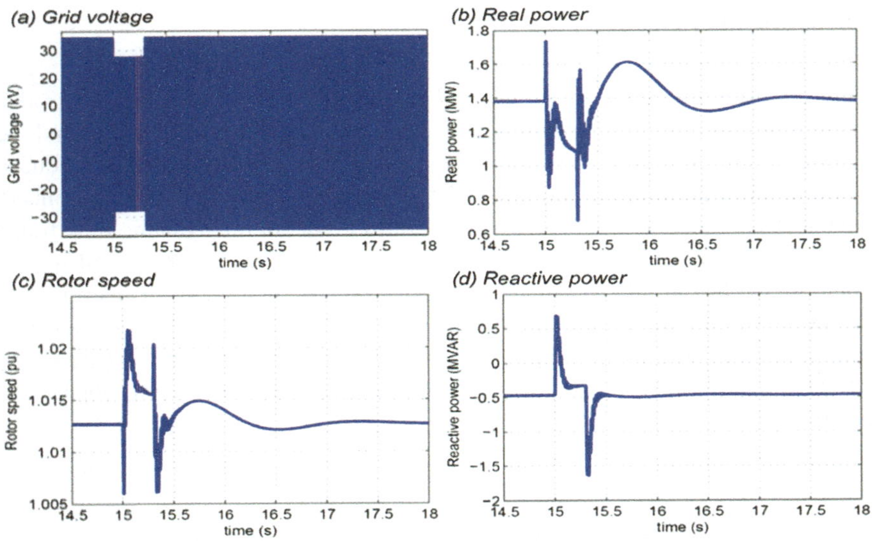

Fig. 10 Type-I FSIG Simulink Results

10.2 Type-II WTs Simulink Results

The changes in grid voltage when the sag begins and ends are expected to result in an immediate answer. The output is greater (about 5 percent) than that of the wind turbine with a set speed. The real power and the reactive power generated by the wind turbines is also disturbed; however, because of the rotor control, the disturbance again differs qualitatively and quantitatively from the response of the fixed wind turbine. The findings also show that after the sag stops a mechanical oscillation occurs, and that the oscillation damps out [19] (Figs. 10 and 11).

10.3 Type-III DFIG WTs Simulink Model and Results

See Figs. 12, 13, 14 and 15.

10.4 Type-IV DFIG WTs Simulink Model and Results

The above simulink results shown in Fig. 16 are comparing the power output during cut-in speed (4 m/s) and cut-out speed (12 m/s).

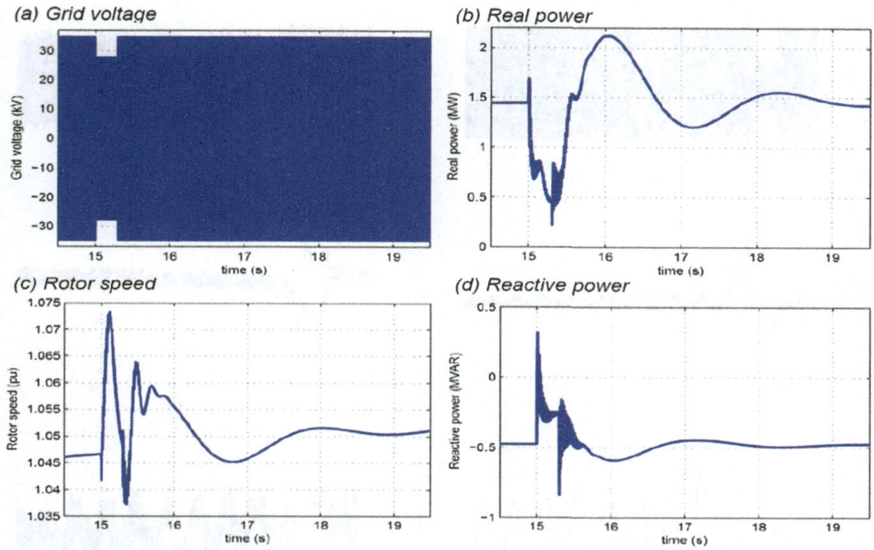

Fig. 11 Type-II DFIG Simulink Results (Grid Voltage, Output Current, Real Power, Reactive Power, Rotor Speed and Pitch Angle)

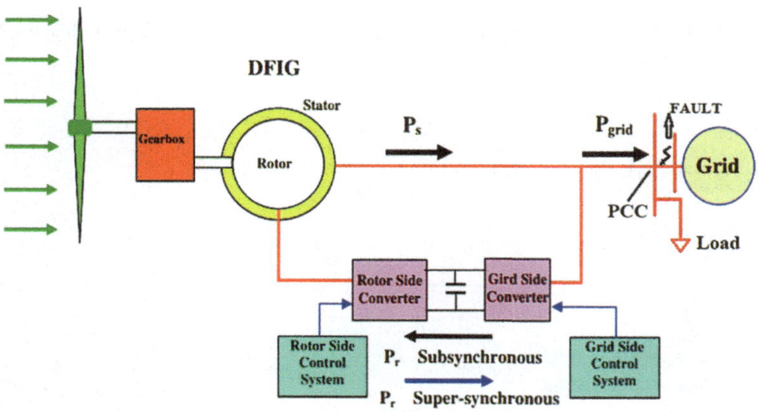

Fig. 12 Type-III DFIG simulink block diagram

11 Discussion

See Tables 2 and 3.

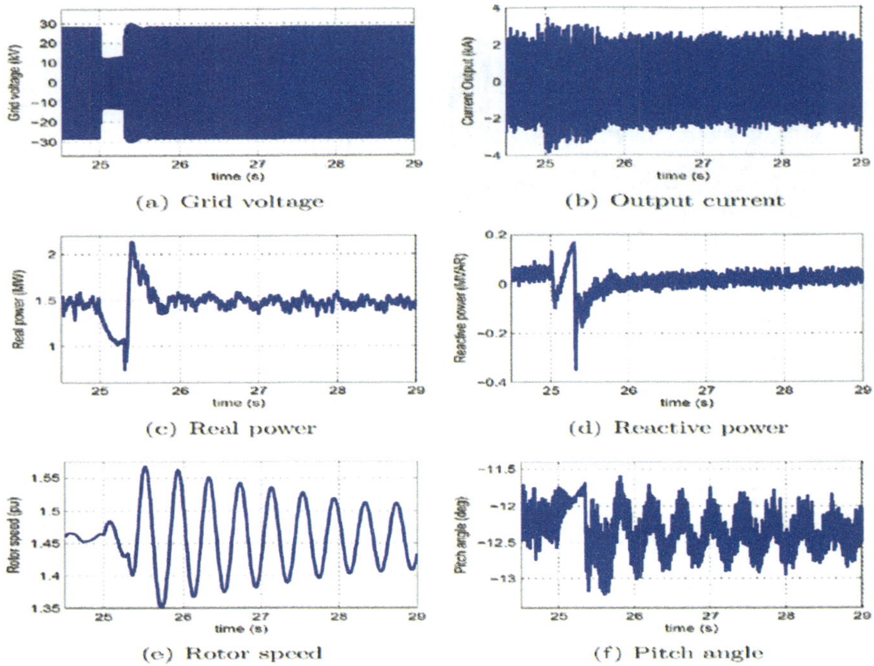

Fig. 13 Type-III DFIG simulink results (Grid Voltage, Output Current, Real Power, Reactive Power, Rotor Speed and Pitch Angle)

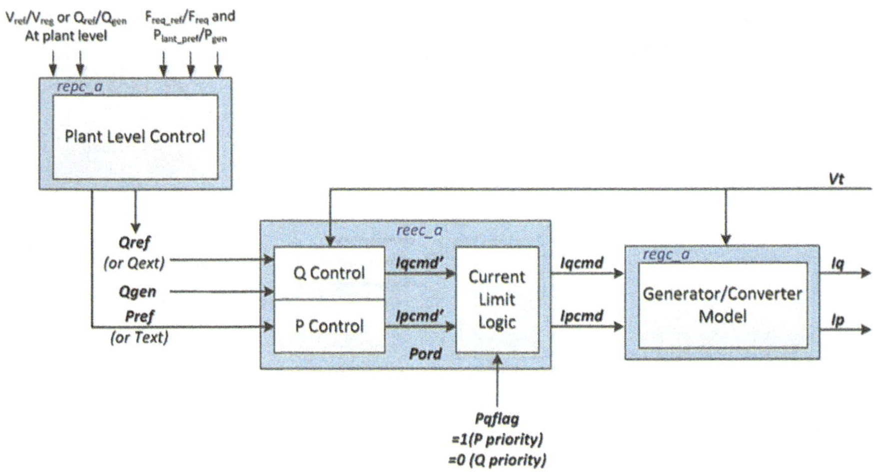

Fig. 14 Type-IV PMSG control design [20]

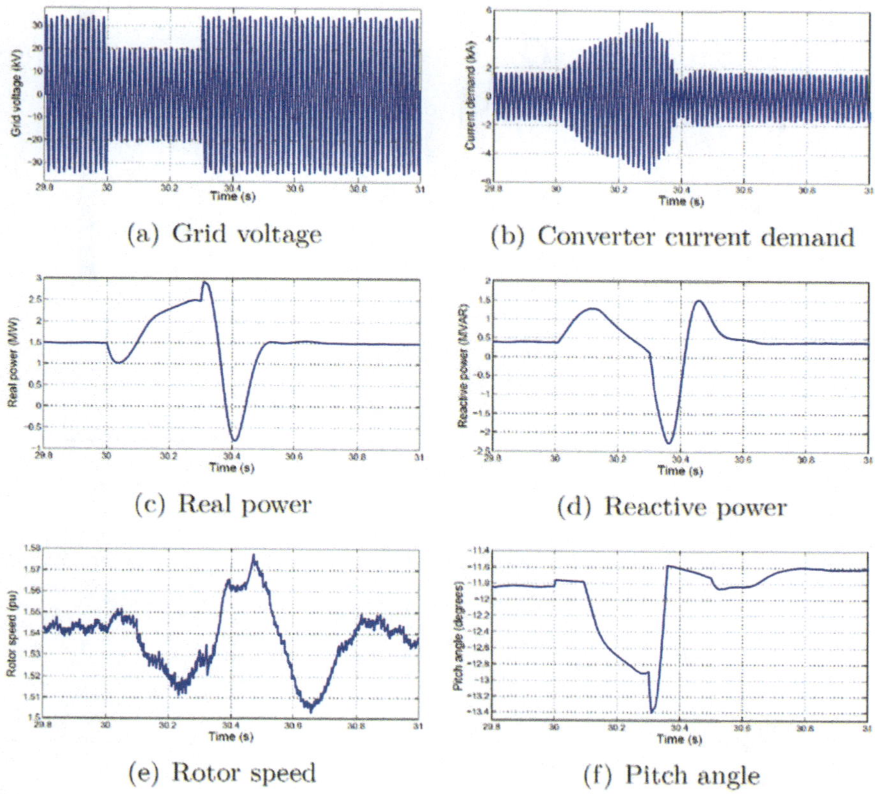

(a) Grid voltage (b) Converter current demand (c) Real power (d) Reactive power (e) Rotor speed (f) Pitch angle

Fig. 15 Type-IV PMSG Simulink Results

12 Conclusion

From the simulink design and electro-mechanical transformation conclude the DFIG and PMSG can be use according to the demand of the grid and cost of the farm shown in Tables 2 and 3. Wind power is being used worldwide very quickly and wind turbines are using many different technologies. This paper evaluates the three main types of wind turbines used over the last 20 years. At the load end of a transmission line connected to the wind generators demands the faster response towards the wind speed variations and full control over active and reactive powers. Hence DFIG and PMSG based Type-III and Type-IV wind turbine system installed in the grid integrations for onshore and offshore applications to facilitate the load demands.

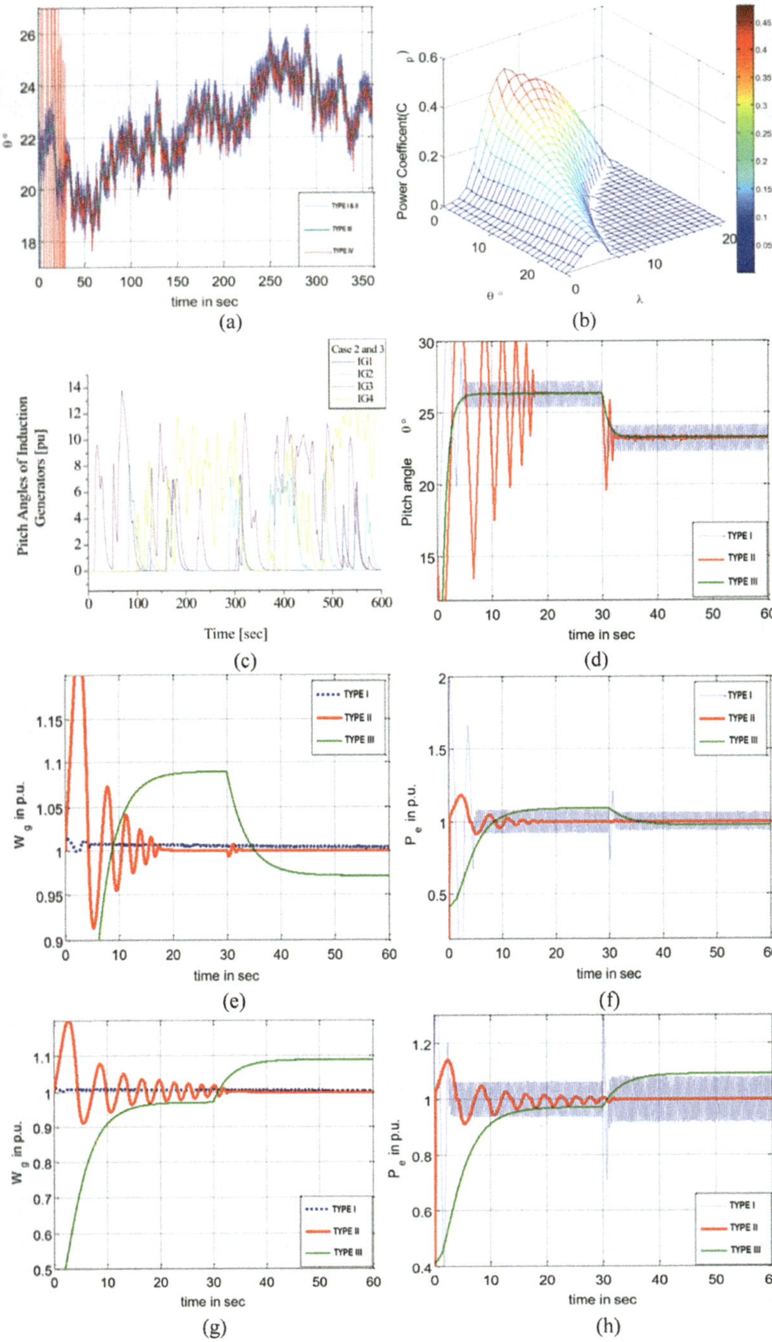

Fig. 16 Comparison based on Simulink Results **a** Pitch Angle generation **b** Power Coefficient generation **c** Pitch Angle generation (pu) **d** Pitch Angle variation **e** Generated Active Power during 4 m/sec **f** Generated Reactive Power during 4 m/sec **g** Generated Active Power during 12 m/sec **h** Generated Reactive Power during 12 m/sec

Table 2 Types of wind turbine system comparison (type based)

Indices	Type I Wind Turbine	Type II Wind Turbine	Type III Wind Turbine	Type IV Wind Turbine
WT Mode	Fixed Speed	Variable Speed Partial control	Variable speed Full control	Variable speed Full control
Generator	SCIG	WRIG	DFIG	PMSG
Grid Link	Soft starter Capacitor bank	Partial scale converter	Full scale converter	Full scale converter
Mechanical Drive	Gear	Gear	Gear	Gear less
Speed Range	~2% Slip N_{synch}	$-30 \pm 30\%$ N_{synch}	0-1 N_{rated}	0-1 N_{rated}
PAC (Pitch Angle)	Fixed	PAC control	PAC control	PAC control
Controller	Rotor	RSC and PAC	PAC, RSC, GSC	PAC, RSC, GSC
Power Curve	Minimal	Average	Fully controlled	Fully controlled

Table 3 Types of wind turbine system comparison (technique based)

Performance	Machine Types and Techniques	
	Type-III DFIG based WTs	Type-IV PMSG based WTs
Manufacturing	Complicated Rotor, Simple Stator	Simple Rotor, Simple Stator
Stator	Radial, Coreless	Radial, Coreless
Rotor	Coil	Permanent Magnet
Carbon Brush	Inbuilt	Not Required
Slip Ring	Inbuilt	Not Required
Converter	30-35% Output	100% Output
LVRT	Needed	Needed
Reactive Power	Demand Controller	Fully Controlled
Grid Support	Average/Good	Very Good
Maintenance	Highly Needed	Very Less
Cost	High Maintenance Cost	High Initial and Installation Cost
Advantages	• Low initial investment • Fast Response • Heavy load with maintenance • Medium/High efficiency • High speed needs a gearbox • Cheaper than PMSG + Full converter	• No coil and No slip rings • No rotor maintenance • Low, Medium and High speed use • High Efficiency • Small and Lighter generator • High Annual Power
Disadvantages	• Difficult to service the rotor hub • No Low and medium speed operation • Low efficiency • Needs more controller for smooth and high efficiency	• High initial investment • Demand high standard software • Sophisticated Process

References

1. B.P. Ganthia, S.K. Barik, Steady-state and dynamic comparative analysis of PI and fuzzy logic controller in stator voltage oriented controlled DFIG fed wind energy conversion system. J. Inst. Eng. India Ser. B **101**, 273–286 (2020). https://doi.org/10.1007/s40031-020-00455-8
2. B.P. Ganthia, S.K. Barik, B. Nayak, Shunt connected FACTS devices for LVRT capability enhancement in WECS. Eng. Technol. Appl. Sci. Res. **10**(3), 5819–5823 (2020)
3. B.P. Ganthia, Application of hybrid facts devices in DFIG based wind energy system for LVRT capability enhancements. J. Mech. Cont. Math. Sci. **15**(6), 245–256 (2020). https://doi.org/10.26782/jmcms.2020.06.00019
4. B.P. Ganthia, Application of hybrid facts devices in DFIG based wind energy system for LVRT capability enhancements. J. Mech. Cont. Math. Sci. **15**(6), 139–157 (2020). https://doi.org/10.26782/jmcms.2020.06.00012
5. A. Honrubia-Escribano, F. Jiménez-Buendía, E. Gómez-Lázaro, J. Fortmann, Field validation of a standard type 3 wind turbine model for power system stability, according to the requirements imposed by IEC 61400-27-1. IEEE Trans. Energy Convers. **33**(1), 137–145 (2018)
6. L. Sun, C. Peng, J. Hu, Y. Hou, Application of type 3 wind turbines for system restoration. IEEE Trans. Power Syst. **33**(3), 3040–3051 (2018)
7. E. Guest, K.H. Jensen, T.W. Rasmussen, Sequence domain harmonic modeling of type-IV wind turbines. IEEE Trans. Power Electron. **33**(6), 4934–4943 (2018)
8. J. Hu, L. Sun, X. Yuan, S. Wang, Y. Chi, Modeling of type 3 wind turbines with df/dt inertia control for system frequency response study. IEEE Trans. Power Syst. **32**(4), 2799–2809 (2017)
9. J. Jallad, S. Mekhilef, H. Mokhlis, Frequency regulation strategies in grid integrated offshore wind turbines via VSC-HVDC technology: a review. Energies **10**, 1244 (2017)
10. H.S. Ko, G.G. Yoon, N.H. Kyung, W.P. Hong, Modeling and control of DFIG-based variable-speed wind-turbine. Electric Power Syst. Res. **78**(11), 1841–1849 (2008), ISSN 0378-7796, https://doi.org/10.1016/j.epsr.2008.02.018
11. B.P. Ganthia, V. Agarwal, K. Rout, M.K. Pardhe, Optimal control study in DFIG based wind energy conversion system using PI & GA, in *International Conference on Power and Embedded Drive Control (ICPEDC)* (Chennai, 2017), pp. 343–347
12. B.P. Ganthia, S. Mohanty, P.K. Rana, P.K. Sahu, Compensation of voltage sag using DVR with PI controller, in *2016 International Conference on Electrical, Electronics, and Optimization Techniques (ICEEOT)* (Chennai, 2016), pp. 2138–2142. https://doi.org/10.1109/iceeot.2016.7755068
13. B.P. Ganthia, P.K. Rana, T. Patra, R. Pradhan, R. Sahu, Design and analysis of gravitational search algorithm based TCSC controller in power system. Mater. Today Proc. **5**(1), 841–847 (2018)
14. A. Moghadasi, A. Sarwat, J.M. Guerrero, A comprehensive review of low-voltage ride through methods for fixed speed wind power generators. Renew. Sustain. Energy Rev. **55**, 823 –839https://doi.org/10.1016/j.rser.2015.11.020
15. K. Siraj, H. Siraj, M. Nasir, Modeling and control of a doubly fed induction generator for grid integrated wind turbine, in *2014 16th International on Power Electronics and Motion Control Conference and Exposition (PEMC)* (IEEE, 2014)
16. H.W. Kim, S.S. Kim, H.S. Ko, Modeling and control of PMSG-based variable-speed wind turbine. Electric Power Syst. Res. **80**(1):46–52 (2010), ISSN 0378-7796, https://doi.org/10.1016/j.epsr.2009.08.003
17. NREL, Dynamic Models for Wind Turbines and Wind Power Plants (NREL/SR-5500-52780) (2011)
18. M. Singh, M. Vyas, S. Santoso, Using generic wind turbine models to compare inertial response of wind turbine technologies, in *IEEE Power and Energy Society General Meeting*, (Minneapolis, MN, 2010)

19. A. Perdana, O. Carlson, Aggregated models of a large wind farm consisting of variable speed wind turbines for power system stability studies, in *Proceedings of 8th International Workshop on Large-Scale Integration of Wind Power into Power Systems as well as on Transmission Networks for Offshore Wind Farms* (Bremen, Germany, 2009)
20. J. Manwell, J. McGowan, *Wind Energy Explained: Theory, Design And Application.* Wiley Chichester (2003)

Optimal Planning of Reactive Power in Power Transmission System Ensuring System Security Using Probabilistic-CSAJAYA

Nihar Karmakar and Biplab Bhattacharyya

List of Symbols

$V_{i,}\ V_j$	ith and jth bus voltages
G_k	Line Conductance
$\delta_{0i,}\ \delta_{0j}$	ith and jth bus phase angles respectively
$\mathbb{P}_{Gi,}\mathbb{Q}_{Gi}$	Active and VAR generation at the ith bus
$\mathbb{P}_{Di,}\mathbb{Q}_{Di}$	Active and VAR demand at the ith bus
$B^l\ G^l$	Line susceptance and transfer conductance
N_{ch}	Line charging elements
Y_{ch}	Admittance of line charging
$\mathbb{Q}_{ch}$	Total reactive power supplied during charging
$\mathrm{n_g}$	Number of PV buses
$\mathrm{n_b}$	Total number of buses
V_G	Generator bus voltage
V_G^{max}	Maximum limit of generator voltage
V_G^{min}	Minimum limit of generator voltage
$\mathbb{Q}_G$	Generated reactive power
$\mathbb{Q}_G^{max}$	High limit of generated reactive power
$\mathbb{Q}_G^{min}$	Low limit of generated reactive power
$OLTC$	Open loop tap changing transformers
tap	Tap settings of OLTC
m	Number of lines

N. Karmakar (✉)
Department of Electrical Engineering, Indian Institute of Technology (ISM), Dhanbad 826004, India
e-mail: nihar.16dr000115@ee.ism.ac.in

B. Bhattacharyya
Department of EE, IIT (ISM), Dhanbad 826004, India
e-mail: biplab@iitism.ac.in

V. K. Singh et al. (eds.), *Renewable Energy and Future Power Systems*, Energy Systems in Electrical Engineering, https://doi.org/10.1007/978-981-33-6753-1_10

n_{tap}	Number of transformer
Y_{ii}	Sending end admittance
Y_{jj}	Receiving end admittance
n_{svc}	Number of SVCs
pop	Population
B_{svc}	Susceptance of SVC
y_{svc}	Admittance of SVC

1 Introduction

Modern power grids of several countries have been experiencing sudden blackout due to reactive power [1, 2] from decades which create more incitement toward secure operation and planning. With the ever-increasing load demands, transmission system has been suffering from voltage instability and losses. So, optimal reactive power planning (RPP) is the means of deliverance from voltage instability and losses.

Extensive literature survey has reported a number of specific descriptions of RPP from decades. It is illustrated by the following actions:

- Installation and sizing of VAr sources optimally [3].
- Minimization of real power losses [4] and operating cost [5].
- Optimal management of VAr reserves [6].
- Allocation of multi-type FACTS devices [7].
- Optimal planning of reactive power in photovoltaic [8] wind energy integrated power system [9].
- Utilization of power electronic converter to control the generation and absorption of Q-power [10].
- Exact distribution of remuneration [11] and procurement of reactive power [12] in deregulated system.
- Enhancement of security of systems through voltage stability [13].

So, RPP is a deterministic complex job for the power system planners to address the above means at a time. It is also noteworthy that RPP researchers are facing teething difficulties such as determination of optimal parameter settings within their limits, operation of transmission system with minimum energy loss, and proper installation of additional VAr compensators to the system.

The continuous variables such as shunt and series compensators, transformer tapping ratio make RPP as a deterministic nonlinear problem. Authors in [14] used interval Taylor expansion method to determine the state variables and through linear approximation method for the optimization of VAr. RPP through the installation of slow and fast VAr compensators is achieved by using particle swarm optimization (PSO) in [15]. For the handling of power flow constraints, the authors implemented enhanced PSO [16] to decrease voltage deviation along with losses. To solve the model of reactive power optimization genetic algorithm (GA) used in [17]. The article

[18] proposed hybridization between PSO and GA to eliminate the stagnation of controlling variables subject to power loss minimization. An effort using differential evolutionary (DE) algorithm has been seen in [19] toward the searching of pareto-optimal solutions of RPP and voltage instability. In [20] fuzzy-TLBO-based strategy was rendered to determine the optimal variables for optimal reactive power flow. Population-based evolutionary algorithms were used to remove the sizing riddles of series FACTS device in Indian power system [21]. Authors in [22] implemented algorithms to compute the VAr sensitivity along with the improvement of voltage instability.

1.1 Aims and Contribution

It is seen from the literature review that most of the research article primarily focused upon operational aspects of RPP rather than economic security. So, the authors are primarily showing their incitement to minimize the system operating cost. Therefore, this article aims to address the technical issues subject to RPP with detail analysis of solution methodology, optimization, and economical aspects of the proposed strategy. The code profiling for optimal solution is very complex due to the involvement of several design variables. These parameters apprize the complexity because of their mutual interaction and influence directly on the cost components. So, this research work has been focused on the technology for the optimal planning of VAr, minimization of losses, and maximization of cost-effective safety.

For the shooting of targets the following contributions have been made in this article.

- Formation of single-objective function with proper co-ordination of different cost components.
- Abrogation of bottlenecks related to size and amount prior to the installation of FACTS devices.
- Facilitation of complexity of different controlling parameters.
- Utilization of the recent algorithms and their hybridization.
- Sharp and minute analysis of the results to establish the effectiveness of the proposed approach.

1.2 Paper Schemes

The remaining portions of this article are presented as follows: Sect. 2 describes the establishment of objective function. Approaches of the proposed methodology are narrated in Sect. 3. Section 4 describes the results of both systems with discussions. Finally, the conclusion is given in Sect. 5.

2 Establishment of Objective Function

The objective of proposed RPP strategy is to optimize the operating cost (OC) ensuring system security. The possible cost components/effects of VAr are included in objective function (OF). The numerical expression of OC is expressed in (1).

$$O.C_{total} = C_{Crp} + CCsvc + CCqg + CCch \tag{1}$$

Here, the total OC is an amalgamation of four different cost components which are stated below. Here the annual cost, considering long-term planning, is taken for the calculation purpose.

Cost component corresponding to real (active) power loss, C_{Crp}

The cost incurred due to P_L is calculated as

$$\begin{aligned} C_{Crp} &= P_L \times rate\,of \\ &= P_L \times (0.06 \times 10^5 \times 24 \times 365) \end{aligned} \tag{2}$$

While energy rate = 0.06$/kWh for 24 h and 365 days.

Numerical formula of real power loss, P_L of a power network is

$$P_L = \sum_{k=(i,j)}^{m} g_k\left[V_i^2 + V_j^2 - 2V_iV_j\text{Cos}(\delta_i - \delta_j)\right] \tag{3}$$

Cost associated with FACTS devices, C_{FACTS}

Here C_{FACTS} is blended with two cost components C_{SVC} (cost due to SVC) and C_{TCSC} (cost due to TCSC). The expense of FACTS devices (C_{FACTS}) may be calculated according to Siemens AG database as:

$$C_{FACTS} = \propto Q2 + \beta Q + \gamma \tag{4}$$

Q represents the reactive power support from different types of FACTS devices in MVAR. The cost coefficient values [23] of different FACTS devices (α, β, and γ) are tabulated in Table 1.

Reactive generator power cost, C_{Cqg}

The expression for C_{Cqg} is

$$\begin{aligned} C_{Cqg} &= Rate\,of\;Q\_generation\,cost \times \mathbb{Q}_G \\ &= (0.0068 \times 24 \times 365) \times \mathbb{Q}_G \end{aligned} \tag{5}$$

$$\mathbb{Q}_G = \sum_{d=1}^{ng} \left[a_q \mathbb{Q}^2(d) + b_q \mathbb{Q}(d) + c_q\right] \tag{6}$$

Line charging cost, C_{Cch}

From literature [24] we are familiarized that the reactive power cost during line charging by line charging reactances also plays an unavoidable role in planning problem. In RPP it is a practice that if any VAr consumers or accessories identified, then we must consider the expense of that elements in objective function. So we proposed to include this reactive power source as a part in the VAr contribution calculations. Therefore,

$$\begin{aligned} C_{Cch} &= \mathbb{Q}_{ch} \times VAr\;p.u\,cost\,during\,line\,charging \\ &= \mathbb{Q}_{ch} \times 11.6068 \end{aligned} \tag{7}$$

Toward any branch (i-j), VAr supplied at the moment of line charging can be measured as:

Table 1 FACTS device cost-coefficient values

Coefficients	FACTS devices	
	SVC	*TCSC*
α	0.0003	0.0015
B	−0.3051	−0.7130
Γ	127.38	153.75

$$\left.\begin{aligned} Q_{c_{i-j}} &= V_i^2 \tfrac{Y_{ch}}{2} + V_j^2 \tfrac{Y_{ch}}{2} \\ \mathbb{Q}_{ch} &= \sum_{ch=1}^{N_{ch}} Q_{c_{i-j}} \end{aligned}\right\} \quad (8)$$

The values of the variables in Eqs. (2), (5) and (7) were collected from [25]. All rates are taken in $ but it can be easily converted in Indian currency (Rs.) by currency exchange services.

So in a nut shell, the conclusive objective function of this proposed article is

$$\text{Minimize, } OC_{total} = \sum C_{Crp} + \left(\sum_{c=1}^{n_c} C_{Csvc} + \sum_{g=1}^{ng} C_{Cqg} + \sum_{ch=1}^{N_{ch}} C_{Cch} \right) \quad (9)$$

Eq. (9) shows its additive nature toward the readers. Hence, indeed, it may be considered as a split objective function in place of a multi-objective one.

2.1 System Constraints

Primarily Eq. (9) is solved stepwise considering two types of constraints.

2.1.1 Equality Constraints

During load flow analysis these constraints should be maintained. The balanced constraints are formulated in Eqs. (10) and (11).

$$\mathbb{P}_{Gi} - \mathbb{P}_{Di} - V_i \sum_{j=1}^{n_b} V_j \left[G_{ij}^l \cos(\delta_{0i} - \delta_{0j}) + B_{ij}^l \sin(\delta_{0i} - \delta_{0j}) \right] = 0 \text{ where i } = 1, 2, .., \text{nb} \quad (10)$$

$$\mathbb{Q}_{Gi} - \mathbb{Q}_{Di} - V_i \sum_{j=1}^{n_b} V_j \left[G_{ij}^l \sin(\delta_{0i} - \delta_{0j}) - B_{ij}^l \cos(\delta_{0i} - \delta_{0j}) \right] = 0 \text{ where i } = 1, 2, .., \text{nb} \quad (11)$$

2.1.2 Inequality Constraints

The inequality constraints for the above-mentioned problem are shown in Eq. (12)

$$\left.\begin{array}{r} V_{Gi}^{\min} \le V_{Gi} \le V_{Gi}^{\max};\ \ i = 1, 2.., nb \\ Q_{Gi}^{\min} \le Q_{Gi} \le Q_{Gi}^{\max};\ \ i = 1, 2, .., nb \\ tap_i^{\min} \le tap_i \le tap_i^{\max};\ \ i = 1, 2, .., n_{tap} \\ SVC_i^{\min} \le SVC_i \le SVC_i^{\max};\ \ i = 1, 2, .., n_{svc} \\ TCSC_i^{\min} \le TCSC_i \le TCSC_i^{\max};\ \ i = 1, 2, .., n_{t\,csc} \end{array}\right\} \quad (12)$$

2.2 *Influence of Tap Settings*

Basically, tap settings of OLTC play an important role in optimal power flow through Y_{bus} matrix modification. The effect of tap positions is described mathematically below. Suppose we have a two-bus transmission network (Fig. 1).

$$I_i = \frac{y_{ij}}{tap}(V_{sec} - V_j)$$

The current running from bus i,

$$or,\ I_i = \frac{y_{ij}}{tap}\left(\frac{V_i}{tap} - V_j\right) = \frac{y_{ij}}{tap^2}(V_i - tapV_j) \quad (13)$$

$$I_j = \frac{y_{ij}}{tap}(V_i - V_{sec})$$

The current running from bus j,

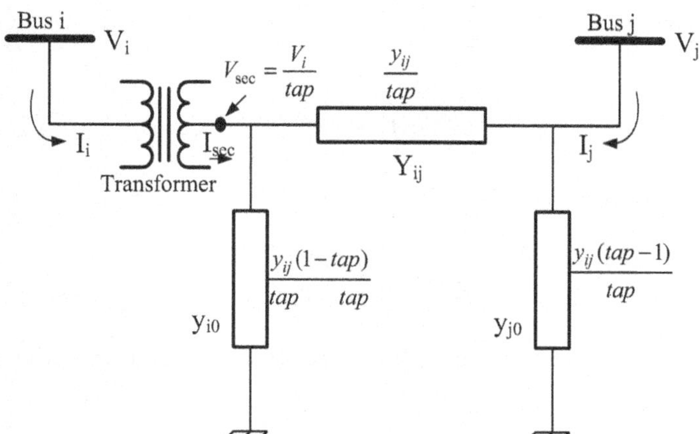

Fig. 1 Equivalent circuit diagram with actual values of admittance

$$or,\ I_j = y_{ij}\left(V_j - \frac{V_i}{tap}\right) = \frac{y_{ij}}{tap}\left(tapV_j - V_i\right) \tag{14}$$

In the above diagram, the p.u. transformer admittance both series and shunt is considered while they are the reciprocal of p.u. impedance of the transformer. From the experience of KCL on node i and j neglecting shunt admittance,

$$I_i = Y_{ij}(V_i - V_j) + y_{i0}V_i \tag{15}$$

$$I_j = Y_{ij}(V_j - V_i) + y_{j0}V_j \tag{16}$$

where Y_{ij} *(new line admittance) = line admittance* (y_{ij}) + *Transformer admittance* (y_{i0} *and* y_{j0}).

Now considering $V_i = 0$ and $V_j = 1$, from Eqs. (15) to (13)

$$I_i = -Y_{ij} \tag{17}$$

and

$$I_i = -y_{ij}/tap \tag{18}$$

Comparing (17) and (18),

$$Y_{ij} = y_{ij}/tap \tag{19}$$

Again substituting $V_i = 1$ and $V_j = 0$, from Eqs. (18) to (16)

$$I_j = Y_{ij} + y_{j0} = y_{ij}/tap + y_{j0} \tag{20}$$

and

$$I_j = y_{ij} \tag{21}$$

From Eqs. (20) and (21),

$$y_{j0} = y_{ij}\left(\frac{tap - 1}{tap}\right) \tag{22}$$

Now substituting the value of Y_{ij} from Eqs. (19) to (15) and comparing with (13)

$$y_{i0} = \frac{y_{ij}}{tap}\left(\frac{1 - tap}{tap}\right) \tag{23}$$

So, the elements of Y_{bus} matrix of this system can be considered as:

$$Y_{ii} = \frac{y_{ij}}{tap} + \frac{y_{ij}}{tap}.\frac{1 - tap}{tap} = \frac{y_{ii}}{tap^2} \tag{24}$$

$$Y_{jj} = \frac{y_{ij}}{tap} + \frac{y_{ij}}{tap}.(tap - 1) = y_{ij} \tag{25}$$

Therefore,

$$Y_{ij} = Y_{ji} = -\frac{y_{ij}}{tap} \tag{26}$$

In summary, the modifications of Y_{bus} matrix due to the involvement of transformer are as follows:

1. The sending end bus (or the nearest bus connected to the transformer) self-admittance is replaced by y_{ii}/tap^2 in place of y_{ij}
2. No change in case of self-admittance of receiving end bus
3. Changes happened in mutual admittance of transformer from $-y_{ij}$ to $-y_{ij}/tap$.

3 Approaches of Proposed Methodology

From previous sections it is realized that RPP is very hard to solve directly. Also, the problem formulation is a mixture of nonlinear equations for which global optimized set of controlling parameters can't be achieved by implementing conventional optimization approach only. So, the solution of RPP is obtained by the following sub-strategies:

3.1 Detection of Candidate/Weak Nodes and Lines

For minimizing (4), it will be difficult computationally if all variables (size, location, and switching) related to static VAr compensators (SVC) and thyristor-controlled series compensators (TCSC) are needed to be simultaneously. Hence, strategically, one economical process is to identify the optimal locations for the SVC and TCSC placement which may lead to a measureable power loss reduction prior to solving the RPP problem. Therefore, this research work tried to pick out critical/weak/candidate buses and lines in such a fashion that appropriate measures to be done by the system operators to solve the problem of voltage instability. In this proposed article, the following methods have been taken for weak bus and line detection purpose.

3.1.1 Methods for Bus Detection

- Loss sensitivity analysis (LSA) [26]
- Power flow analysis (PFA): [27]
- Modal analysis (MA): [27]

3.1.2 Methods for Bus Detection

- Voltage collapse proximity indication (VCPI) method [28]
- Power flow analysis (PFA) method
- Fast voltage stability index (FVSI) method [28]

3.2 Modeling of SVC

SVC is a combination of thyristor-controlled reactors along with a fixed capacitor bank. When it is connected parallel in line, the reactive power current is drawn from the sources and the thyristor valves connect or isolate shunt-connected reactors and capacitors. SVC injects VAr to a bus as: $\mathbb{Q}_{svc} = B_{svc}V^2$.. In this proposed work, SVC is installed in this manner:

```
for k= 1: n_svc
    shunt(pop_svc(k)) = sqrt(-1)*svc_value
end
```

n_{svc} and *svc_value* is number of SVC units and p.u. susceptance value of SVC unit, respectively. After incorporating SVC units, the bus admittance matrix (Y_{bus}), used for load flow analysis, is reformed as:

$$Y_{bus}^{svc} = Y_{bus} + \begin{bmatrix} 0\,0\ldots\ldots0\,0 \\ 0\;y_{svc}\ldots0\,0 \\ \ldots\ldots\ldots\ldots \\ 0\,0\ldots\ldots0\,0 \\ 0\,0\ldots\ldots0\,0 \end{bmatrix}_{i\times j}$$

By adjusting the symmetry and sparsity of this bus admittance matrix, it is possible to lower the required mathematical calculations related to system power loss.

3.3 Application/Adaptation of Evolutionary Algorithms

After addressing technical aspects, some well-established evolutionary algorithms such as PSO, DE, sine–cosine algorithm (SCA) are used to search an optimal set of controlling parameters prior to solving the previously stated objective function. Inspired from "No free lunch theorem" [29], no single optimizer exists which is able to determine a consistence solution to all global optimization problem, and the authors adapt hybrid optimization method for RPP problem. A unique hybridization between crow search algorithm (CSA) and JAYA is simulated on standard power system for VAr planning purpose. Along with this, in this proposed article, some other algorithms are also applied to determine the optimized set of controlling parameters. In later section the hybridization is described briefly along with CSA and JAYA.

3.3.1 Crow Search Algorithm [30]

Indeed, CSA is a population-based algorithm used to determine the set of controlling parameters within their limits following the intelligent behavior of crows for finding, storing, and retrieving their food in environment. In CSA, every position of a particular crow represents a most feasible solution of an optimization-oriented problem. At initialization the position vector of crows is represented in a string by ${}^{0}X^{i,iter} = [x_1^{i,iter}, x_2^{i,iter}, \ldots, x_d^{i,iter}]$ while *iter* = *1,2, …, max_iter*. These memory locations of the ith crow, best food location, are expressed as ${}^{0}M^{i,iter} = \left[m_1^{i,iter}, m_2^{i,iter}, \ldots, m_d^{i,iter}\right]$. Now, at every iteration ith crow also moves randomly to find the better sources of food (hidden places of other crows) by following the randomly selected jth crow and then acquired the best hiding place $({}^{0}M^{j,iter})$ corresponding to jth crow.

Considering some numerical possibilities, the up-to-date position and hidden place (memory) of ith crow can be determined by (27) and (28).

$$ {}^{0}X^{i,iter+1} = \begin{cases} {}^{0}X^{i,iter} + rand \times fl^{i,iter} \times ({}^{0}M^{j,iter} - {}^{0}X^{i,iter}) & if\ r_j \geq AP \\ a\ random\ position & otherwise \end{cases} \tag{27} $$

$$ {}^{0}M^{i,iter+1} = \begin{cases} {}^{0}X^{i,iter+1} & if\ f({}^{0}X^{i,iter+1})\ is\ better\ than\ f({}^{0}M^{i,iter}) \\ {}^{0}M^{i,iter} & otherwise \end{cases} \tag{28} $$

3.3.2 Traditional JAYA

In the year of 2015, R. V. Rao presented novel JAYA algorithm in [31]. It requires only one phase to complete iteration. Only population size and iteration number are needed for the accomplishment of one cycle which makes the algorithm very simple during engineering applications. If we have an objective function *F(x)*, then according to the JAYA algorithm, the new value of the variables will be

$$ {}^{0}X'_{j,k,i} = {}^{0}X_{j,k,i} + {}^{1}r_{j,i}\left({}^{0}X_{j,best,i} - \left|X_{j,k,i}\right|\right) - {}^{2}r_{j,i}\left({}^{0}X_{j,worst,i} - \left|{}^{0}X_{j,k,i}\right|\right) \quad (29) $$

After computing the necessary calculations, this value will be considered as an input for the next iteration.

3.3.3 Hybrid CSAJAYA

It is worth mentioning that the above-stated algorithms are well performed in different engineering optimization. However, CSA has been failed to guarantee the convergence in some cases [32] because of insufficient exploration from their search space. Therefore, modification has been rendered to enhance their efficiency in RPP problem. The core concept of the proposed hybridization is mainly dependent upon CSA. In original CSA it is seen that mainly two different parameters, AP and fl, have direct influence on searching strategy. The probability of awareness is a deciding parameter for Eq. (27). Here, if $r_j < AP$ then the position of crows are updated randomly which makes CSA less efficient in case of convergence. In the proposed CSAJAYA this equation is reformulated to achieve more reliable convergence. So, the revised equation for position update will be as follows:

$$ X^{i,iter+1} = \begin{cases} X^{i,iter} + rand \times fl^{i,iter} \times (M^{j,iter} - X^{i,iter}) & if\ r_j \geq AP \\ X_{j,k,i} + r_{1,j,i}\left(X_{j,best,i} - \left|X_{j,k,i}\right|\right) - r_{2,j,i}\left(X_{j,worst,i} - \left|X_{j,k,i}\right|\right) & otherwise \end{cases} \quad (30) $$

Maintaining balance between exploration and exploitation, in Eq. (30) new position of crows will be updated in every iteration by following JAYA algorithm while r_j is less than probability of awareness. So, readers also can call hybrid CSAJAYA as probabilistic-CSAJAYA. Pseudo-code profiling of CSAJAYA is painted in algorithm 1.

Algorithm 1 Pseudo Code of CSAJAYA

```
Randomly Initialize a flock of N crows X_i
Evaluate the fitness of each crow
Initialize the memory of crows
  While (itert<Maximum_iter)
    for every crow
    Selection of Awareness Probability (AP) and flight length (fl)
    Generate r_j randomly
      if r_j≥ AP
        Up-to-date the position of crows using Eq.(27),
        else
       Use Eq.(29) for position updating
      end if
    end for
Test the feasibility of newly generated positions
Calculate the fitness of all crows
Up-to-date the memory of crows using Eq. (28)
    iter=(iter+1)
  end while
return & save the best crow
```

4 Numerical Results and Discussion

Intel (R) Core i5 (2.9 GHz) processor has been used to execute the simulation code of the proposed work in MATLAB 2013a. For the establishment of efficiency and effectiveness, the proposed work is simulated on IEEE 30 bus and Indian UPSEB 75 bus system. The weak nodes/buses and lines for the placement of SVC and TCSC are shown in Table 2. The specifications of both systems are tabulated in Table 3.

Table 2 Weak nodes and lines of test systems for FACTS devices

Methods	Weak nodes	Methods	Weak lines
IEEE 30 bus test system			
LSA	21, 7, 15, and 30	*VCPI*	29, 4, 41, and 24
PFA	21, 7, 15, and 17	*PFA*	25, 41, 28, and 5
MA	29, 20, 30, and 14	*FVSI*	15, 17, 18, and 19
UPSEB 75 bus test system			
LSA	46, 50, 55, and 49	*VCPI*	40, 38, and 61
PFA	45, 42, 35, and 41	*PFA*	95, 72, and 29
MA	28, 26, 49, and 17	*FVSI*	17, 14, and 18

Table 3 Details of the test systems

System specifications	Test systems			
	IEEE 30 bus		*UPSEB 75 bus*	
	Value	Details	Value	Details
Number of nodes/buses	30	Pai [33]	75	Agrawal et al. [34]
Number of alternators	5	Pai [33]	14	Agrawal et al. [34]
Slack bus	1	At bus 1	1	At bus 1
Number of lines	41	–	98	–
No. of shunt capacitors	2	At bus 10 and 24	1	At bus 42
Open loop tap changing Transformers	4	At branches 11, 12, 14, and 36	24	At branches 1, 2, 3, 4, 5, 6, 7, 8, 9, 10, 11, 12, 13, 14, 15, 16, 17, 18, 19, 20, 21, 22, 23 and 24
Control parameters	16	–	43	–
VAr limits of the generators	Q_{min}, Q_{max}	Pai [33]	Q_{min}, Q_{max}	Archana [35]
P_L and operating cost ($) without RPP	7.11 MW and 4.0377×10^7 $	–	219.9 MW and 1.1560 × 108 $	–

4.1 System 1: IEEE 30 Bus System

Table 4 presents the optimum values of objective function using different algorithms. Among six algorithms CSAJAYA minimizes the operating cost to 3.6169×10^6 with P_L 0.0688. The convergence characteristic of losses is represented in Fig. 2.

4.2 System 2: Indian UPSEB (Utter Pradesh State Electricity Board) 75 Bus System

The optimized values of loss and operating cost are presented in Table 5. From here it is seen that the minimum operating cost is 7.8097×10^7 with loss of 1.4889 p.u in case of CSAJAYA. The best positions for SVC and TCSC are obtained from LSA

Table 4 Optimal value of P_L and operating cost

Components	Algorithms					
	PSO	*DE*	*SCA*	*JAYA*	*CSA*	*CSAJAYA*
P_L *(p.u.)*	0.0692	0.0691	0.0690	0.0690	0.0690	0.0688
O.C ($)	3.6345×10^6	3.6327×10^6	3.6252×10^6	3.6253×10^6	3.6248×10^6	3.6169×10^6

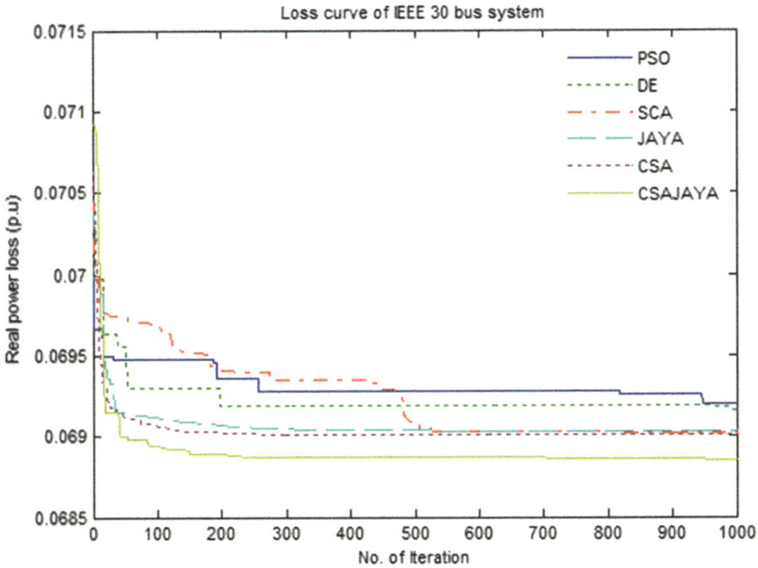

Fig. 2 Convergence characteristics of P_L by different algorithms

and VCPI method. The optimal set of controlling parameters has been tabulated in Table 6.

Figure 3 narrates the convergence characteristics of real power loss. It is seen that probabilistic-CSAJAYA provides a smooth convergence curve after 1000 iterations in comparison to others algorithm. The JAYA and CSA algorithm minimizes P_L to 1.7617 p.u. and 1.5067 p.u., respectively, while they are used individually. But, after hybridization, CSAJAYA provides 1.4889 p.u. loss. Here only the plots of best results are shown.

5 Conclusion

A reactive power planning ensuring security was proposed in this article. A hybridization between two algorithms was rendered to find out the solution of RPP. It was seen that CSA when coupled with JAYA could provide satisfactory solutions of objective function. It was capable to elicit the optimal set of controlling parameters within their ranges related to total operating cost minimization. The smoothness of convergence curves obtained from PSODE algorithm implied the performance of the proposed approach with respect to other soft computing techniques. SVC and TCSC were successfully installed at weak buses and lines by removing technical and computational burdens. These weak positions were determined from different analysis while all constraints were within their limits. At base case scenario the proposed approach was able to reduce both real/active power loss and operational cost satisfactorily.

Table 5 Optimal value of P_L and operating cost

Components	Algorithms					
	PSO	*DE*	*SCA*	*JAYA*	*CSA*	*CSAJAYA*
P_L *(p.u.)*	1.5756	1.8309	1.5111	1.5280	1.5067	1.4889
O.C ($)	8.2806×10^7	9.6226×10^7	7.9410×10^7	8.0009×10^7	7.9180×10^7	7.8097×10^7

Table 6 Optimal system operating variables for system 2

		Methods					
		PSO	DE	SCA	JAYA	CSA	CSAJAYA
Transformer tap settings	*Line no.*	*Tap settings (p.u)*					
	1	1.1	1.0178	1.1	1.1	1.1	1.1
	2	1.1	1.0953	1.1	1.1	1.1	1.1
	3	1.033	1.0062	1.1	1.0016	0.9971	0.9987
	4	1.0976	0.9992	1.1	1.1	1.0933	1.0878
	5	1.1	0.9914	1.1	1.0682	1.1	1.0616
	6	1.0884	1.0348	1.1	1.0245	1.1	1.1
	7	1.031	0.9904	1.1	1.003	1.0809	0.999
	8	1.1	0.9939	1.0738	1.1	1.1	1.0874
	9	1.1	0.9486	1.1	1.0247	1.1	1.0332
	10	1.1	1.0398	1.1	1.1	1.1	1.1
	11	1.1	0.9294	1.0686	1.0171	1.1	1.0988
	12	1.1	0.8977	1.0571	1.1	1.1	1.0802
	13	1.0763	1.0267	1.1	1.1	1.1	1.0969
	14	1.1	0.9421	1.1186	1.0936	1.1	1.1
	15	1.1	1.0064	1.1	1.1	1.1001	1.1
	16	1.1	1.0156	1.1242	1.1	1.1	1.1
	17	1.1	0.9549	1.1187	1.0224	1.0949	1.0226
	18	1.1	1.0537	1.1	1.0016	1.0132	1.0354
	19	1.1	0.9928	1.0997	1.1	1.1	1.1
	20	1.1	1.0299	1.1266	1.1	1.1	1.1
	21	1.1	1.0335	1.1	1.1	1.0984	1.1
	22	1.0715	1.0014	1.039	1.1	1.0537	1.0742
	23	1.1	0.9035	1.1	1.1	1.1	1.0966
	24	1.0006	0.9955	1.0675	0.9895	1.0026	0.9639
SVC at buses	*Bus No.*	*Susceptance value (p.u)*					
	46	0.1500	0.1100	0.0188	0.0330	0.1500	0.1057
	50	0.1144	0.0100	0.1500	0.0100	0.1500	0.1500
	55	0.1500	0.0934	0.1500	0.1494	0.1500	0.0101
TCSC in lines	*Line No.*	*Reactance in (p.u)*					
	40	0.1000	−0.1618	−0.1500	−0.500	0.0200	0.0099
	38	0.1000	−0.2328	0.0400	0.0100	0.0200	0.0100

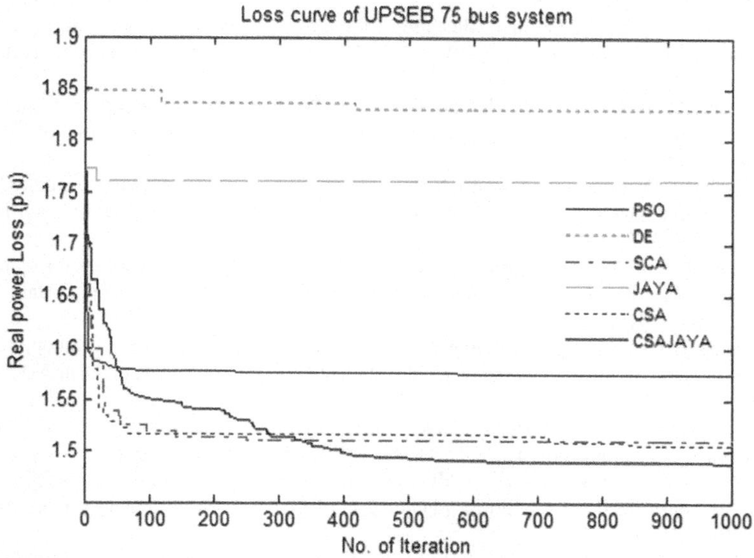

Fig. 3 Convergence characteristics of P_L by different algorithms using LSA and VCPI

References

1. Y.K. Wu, S.M. Chang, Y.L. Hu, Literature review of power system blackouts. Energy Proc. **141**, 428–431 (2017)
2. Y. Jiang, S. Chen, C.C. Liu, W. Sun, X. Luo, S. Liu, D. Forcum, Blackstart capability planning for power system restoration. Int. J. Electr. Power Energy Syst. **86**, 127–137 (2017)
3. A.M. Shaheen, R.A. El-Sehiemy, S.M. Farrag, A novel adequate bi-level reactive power planning strategy. Int. J. Electr. Power Energy Syst. **78**, 897–909 (2016)
4. B. Bhattacharyya, N. Karmakar, A planning strategy for reactive power in power transmission network using soft computing techniques. Int. J. Power Energy Syst. **40** (2020)
5. S. Raj, B. Bhattacharyya, Reactive power planning by opposition-based grey wolf optimization method. Int. Trans. Electric. Energy Syst. **28**, e2551 (2018)
6. B. Bhattacharyya, S. Raj, Differential evolution technique for the optimization of reactive power reserves. J. Circ. Syst. Comput. **26** (2017)
7. C. Li, L. Xiao, Y. Cao, Q. Zhu, B. Fang, Y. Tan, L. Zeng, Optimal allocation of multi-type FACTS devices in power systems based on power flow entropy. J. Mod. Power Syst. Clean Energy **2**, 173–180 (2014)
8. S.S. Alkaabi, H.H. Zeineldin, V. Khadkikar, Short-term reactive power planning to minimize cost of energy losses considering PV systems. IEEE Trans. Smart Grid **10**, 2923–2935 (2018)
9. M. Belkacem, K. Srairi, Solving multi-objective optimal power flow problem considering wind-STATCOM using differential evolution. Front. Energy **7**, 75–89 (2013)
10. An enhanced concept of Q-power management, Bielecki, Sławomir and Tadeusz S. Energy **162**, 335–353 (2018)
11. S.K. Parida, S.C. Srivastava, S.N. Singh, A review on reactive power management in electricity markets. Int. J. Energy Sector Manag. (2011)
12. M. De, S.K. Goswami, Optimal reactive power procurement with voltage stability consideration in deregulated power system. IEEE Trans. Power Syst. **29**, 2078–2086 (2014)

13. K. Kavitha, R. Neela, Optimal allocation of multi-type FACTS devices and its effect in enhancing system security using BBO, WIPSO & PSO. J. Electr. Syst. Inform. Technol. **5**, 777–793 (2018)
14. C. Zhang, H. Chen, Z. Liang, M. Guo, D. Hua, H. Ngan, Reactive power optimization under interval uncertainty by the linear approximation method and its modified method. IEEE Trans. Smart Grid **9**, 4587–4600 (2017)
15. M. Eghbal, N. Yorino, E.E. El-Araby, Y. Zoka, Multi-load level reactive power planning considering slow and fast VAR devices by means of particle swarm optimisation. IET Gener. Transm. Distrib. **2**, 743–751 (2008)
16. G. Chen, L. Liu, Y. Guo, S. Huang, Multi-objective enhanced PSO algorithm for optimizing power losses and voltage deviation in power systems. COMPEL Int. J. Comput. Math. Electr. Electr. Eng. (2016)
17. C. Zhang, H. Chen, H. Ngan, Z. Liang, M. Guo, D. Hua, Solution of reactive power optimisation including interval uncertainty using genetic algorithm. IET Gener. Transm. Distrib. **11**, 3657–3664 (2017)
18. A.A. Esmin, G. Lambert-Torres, A.Z. De Souza, A hybrid particle swarm optimization applied to loss power minimization. IEEE Trans. Power Syst. **20**, 859–866 (2005)
19. J.P. Roselyn, D. Devaraj, S.S. Dash, Multi objective differential evolution approach for voltage stability constrained reactive power planning problem. Int. J. Electr. Power Energy Syst. **59**, 155–165 (2014)
20. A. Moghadam, A.R. Seifi, Fuzzy-TLBO optimal reactive power control variables planning for energy loss minimization. Energy Convers. Manag. **77**, 208–215 (2014)
21. R. Agrawal, S.K. Bharadwaj, D.P. Kothari, Population based evolutionary optimization techniques for optimal allocation and sizing of thyristor controlled series capacitor. J. Electr. Syst. Inform. Technol. **5**, 484–501 (2018)
22. D. Thukaram, K. Parthasarathy, H.P. Khincha, N. Udupa, A. Bansilal, Voltage stability improvement: case studies of Indian power networks. Electr. Power Syst. Res. **44**, 35–44 (1998)
23. S. Ghaemi, F.H. Aghdam, A. Safari, M. Farrokhifar, Stochastic economic analysis of FACTS devices on contingent transmission networks using hybrid biogeography-based optimization. Electr. Eng. **101**, 829–843 (2019)
24. M. De, S.K. Goswami, A direct and simplified approach to power-flow tracing and loss allocation using graph theory. Electric Power Compon. Syst. **38**, 241–259 (2010)
25. J.W. Lamont, J. Fu, Cost analysis of reactive power support. IEEE Trans. Power Syst. **14**, 890–898 (1999)
26. B. Bhattacharyya, S.K. Goswami, R.C. Bansal, Loss sensitivity approach in evolutionary algorithms for reactive power planning. Electric Power Compon. Syst. **37**, 287–299 (2009)
27. N. Karmakar, B. Bhattacharyya, Optimal reactive power planning in power transmission network using sensitivity based bi-level strategy. Sustain. Energy Grids Netw. 100383 (2020)
28. S. Raj, B. Bhattacharyya, Weak bus-oriented optimal Var planning based on grey wolf optimization, in *2016 National Power Systems Conference (NPSC)*, pp. 1–5 (IEEE, 2016)
29. D.H. Wolpert, W.G. Macready, No free lunch theorems for optimization. IEEE Trans. Evol. Comput. **1**, 67–82 (1997)
30. A. Askarzadeh, A novel metaheuristic method for solving constrained engineering optimization problems: crow search algorithm. Comput. Struct. **169**, 1–12 (2016)
31. R. Rao, Jaya: A simple and new optimization algorithm for solving constrained and unconstrained optimization problems. Int. J. Ind. Eng. Comput. **7**, 19–34 (2016)
32. P. Díaz, M. Pérez-Cisneros, E. Cuevas, O. Avalos, J. Gálvez, S. Hinojosa, D. Zaldivar, An improved crow search algorithm applied to energy problems. Energies **11**, 571 (2018)
33. M. A. Pai, *Computer Techniques in Power System Analysis*, 3rd ed. (Tata Mcgraw Hill Education Private Limited, 2005)

34. R. Agrawal, S.K. Bharadwaj, D.P. Kothari, Transmission loss and TCSC cost minimization in power system using particle swarm optimization. Int. J. Innov. Res. Electr. Electron. Instrum. Control Eng. **4**, 226–231 (2016)
35. N. Archana, Exploration of the application of A novel meta-heuristic algorithm for Optimal location of facts devices (Anna University, 2017)

An Architectural and Control Overview of DC-Microgrid for Sustainable Remote Electrification

Yugal Kishor, C. H. Kamesh Rao, R. N. Patel, and Lalit Kumar Sahu

1 Introduction

Increasing use of renewable generation and technical advancement in power electronics converter during the last decade have led to a condition where distributed energy sources (DES) can be integrated easily [1]. Thus, DES along-with other forms of distributed generation (DG) have more potential to cope with present and future energy demand and tend to reduce the consumption of fossil fuels while minimizing the carbon footprints. The DG plays an important role to provide universal electricity access and it is a key to improve the standard of living under the sustainable development goal (SDG) of the United Nations (UN) [2, 3]. However, it is difficult to achieve universal electricity through a conventional way extending the utility grid to remote areas because of limited generation, cost, and power loss associated with transmission and distribution.

Parallelly, solar photovoltaic (SPV) systems are more often for remote electrification either in off-grid [4] or low voltage DcMG [2]. Somehow low power DcMG provides an immediate solution to run very basic loads like lights, mobile charging, and in some cases fans but not a significant as far as improvement in socio-economic uplift is concern [5, 6]. While high power DcMGs are economically unviable [7], therefore, it is important to interconnect the cost-effective nano-grids to run communal loads like hospitals, schools, and water pumping. Many initiatives have been taken for the deployment of off-grid DcMG in developing areas [8]. The government of Bangladesh and India have been set a target to provide rural electrification through the solar home system (SHS) [9–11]. While such standalone systems are available, there is no mechanism to utilize the excess power of these systems. The optimum energy utilization is required to cater the communal load in remote areas [8]. Figure 1 shows the basic architecture of DcMG system, includes the SPV, energy

Y. Kishor (✉) · C. H. Kamesh Rao · R. N. Patel · L. K. Sahu
National Institute of Technology Raipur, Raipur, India
e-mail: yugalsahu87@gmail.com

V. K. Singh et al. (eds.), *Renewable Energy and Future Power Systems*, Energy Systems in Electrical Engineering, https://doi.org/10.1007/978-981-33-6753-1_11

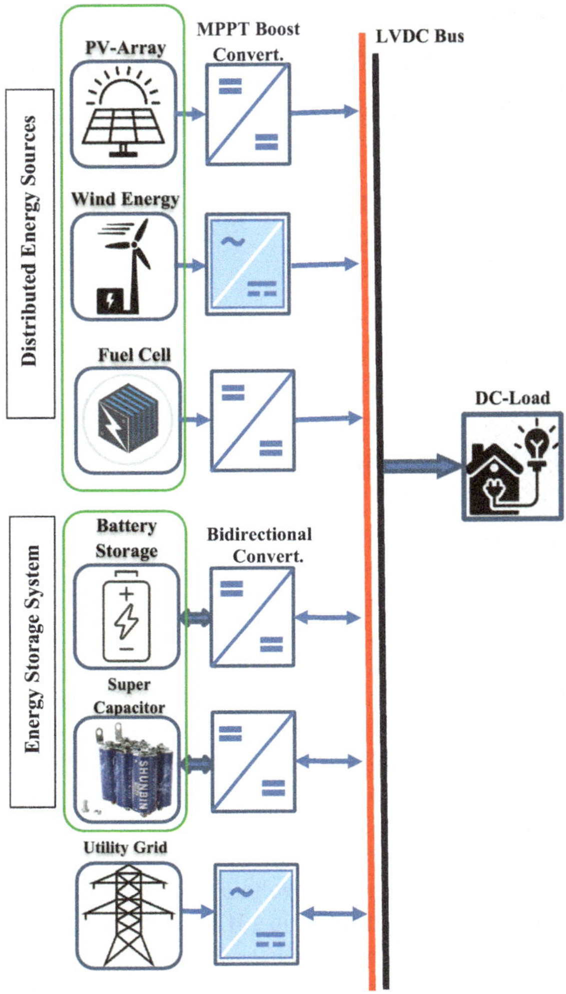

Fig. 1 Typical architecture of DcMG

storage systems (ESS), utility grid, and DC-loads connected to the common DC bus. DcMG can operate either islanded or grid-tie mode depending on load requirement, state of charge (SOC), and solar irradiance level.

In the future grid needs to have following advantages, it should be capable to supply the increasing demand of IoT devices and Electric Vehicles(EV's)which are DC loads [17, 18] and also today power authorities are encouraging the integration of renewable energy sources like Photovoltaic systems, Wind energy, hydro and fuel cell technology [19].

To interface more and more renewable energy sources with the existing power supply grid which results in many advantages like reducing the emission of harmful gases, to minimize the losses which are occurring due conversion from AC to DC and vice versa, digitization, the basic domestic loads include most of the DC systems, which seems like resuming the war between AC and DC, which previously settled in last 1920s [20]. By giving the argument that the transmission cost and losses in AC systems are very less as compared to the DC systems. But as the DC loads are increasing and we have portable DC generation sources like photovoltaic systems and DC ESS, in the 21st century.it is the right time to move towards DC supply systems [21–23]. In this chapter, the importance of remote electrification is discussed in Sect. 2. In Sect. 3, an overview of architecture and comparison of control scheme are reviewed while Sect. 4 highlights the simulation setup. The effectiveness of simulation results are covered in Sects. 5 and 6 suggests the current research aspects in DcMG. The final conclusion of the chapter is discussed in Sect. 7.

2 Importance of Remote Electrification

The reliable access of electricity and its consumption pattern are the figure of merit to measure the socioeconomic standing of any community. The social resource development is directly linked with significant availability of electricity. The unavailability of electricity hampers the human rights to access the clean drinking water, public health delivery, basic education facilities, and suitable lighting, consequently supplementing the poverty and significantly affects the quality of life [12]. There are many adverse effects of unavailability of electricity in remote schools.

The inaccessible basic amenity of lighting, heating, and ventilation in the classroom deteriorates basic education. Moreover, the absence of computing and printing facility, which is considered as a key to modern learning, due to the unavailability of proper electricity in those areas. The children lost interest in schooling and many of them left the school at the primary level itself. Which further led to unemployment and many social issues. The lack of access to electricity may deteriorate the availability of clean water which directly affects the health and sanitation in remote areas. The medical tools are inoperative due to the unavailability of electricity in health care centers, which raises the mortality rate. The provision of electricity access to those regions could help to improve health, education and improved the standard of life and also contribute to catering the potential of agriculture, industrial and employment opportunities [12]. Additionally, green energy remote electrification will further lead to reduce climate change by minimizing the carbon footprints in these regions and control the deforestation rates. In addition to social welfare, there is a huge business potential in the energy market.

3 Overview of Architecture

In the previous decade, many people have access to electricity mostly led by extension of utility grid and interconnection of remote areas, while more dependency on fossil

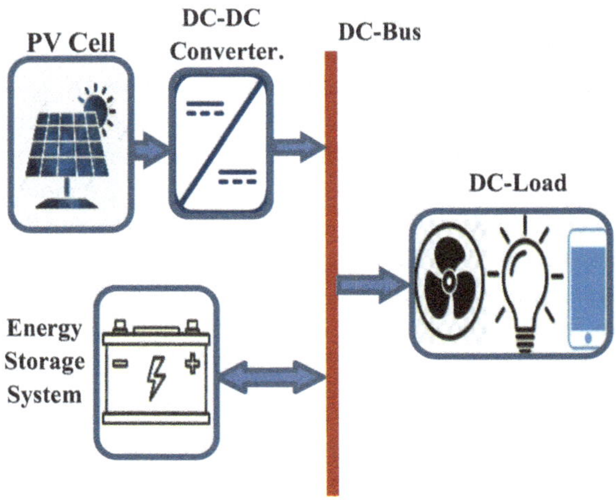

Fig. 2 Typical architecture of SHSS

fuels. The extension of the utility grid has economic constraints due to heavy investment in power projects and transmission line setup. The above issues led to more emphasis on eco-friendly distributed energy resources for remote electrification and due to factors, like rapid depletion of fossil fuels, more awareness about the impact of fossil fuels on the environment, affordable renewable energy technology, and cost-effective MG solution. As most of the Indian regions have significance irradiance for about 300 days, makes it suitable off-grid SPV installations. There are many DcMG architectures are found in the literature [13]. Solar energy adoption for off-grid electrification has grown from stand-alone solar home (SHSS) systems to MG-based deployment [8]. The typical structure of SHSS depicted in Fig. 2, have been deployed in many developing regions of Chhattisgarh, Uttar Pradesh, Andaman Nikobar, and Jharkhand [14, 15]. The basic objective of SHSS is to provide basic electricity access to remote residents. These structures are very efficient as generation, distribution, and utilization of energy all are in dc form, and there are no conversion losses are associated with ac-dc and vice versa conversions. The typical SHSS is capable enough to deliver power to one or more LED bulbs, also able to charge mobile and run a small dc fan. Although SHSS is cost-effective and easy to deploy, it is suboptimal. The advantage of energy-usage diversity can be taken only with resource sharing. Besides, due to the unavailability of a resource-sharing function, these schemes have minimal electrification capabilities and are not feasible for anything demanding, such as water filtration plants/irrigation pumps, school computing loads or village health care units. Figure 3 shows the topological diagram of a typical centralized DcMG system used by many implementations of rural electrification. Such an architecture in which resources for generation (PV panels) and storage (batteries) are located in a central location is referred to as central storage architecture (CGCSA). CGCSA has a unidirectional power flow from a centralized point with SPV generation and storage. There are two dc-dc converter are in this architecture, one for MPPT tracking of SPV and boosting the voltage for distribution while second is required to run the

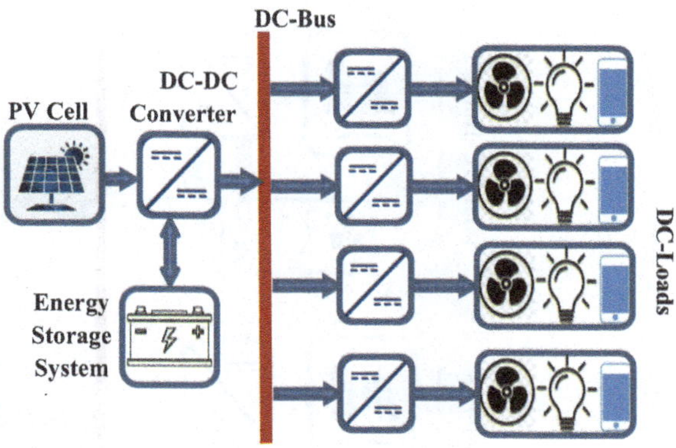

Fig. 3 Typical architecture of CGCSA

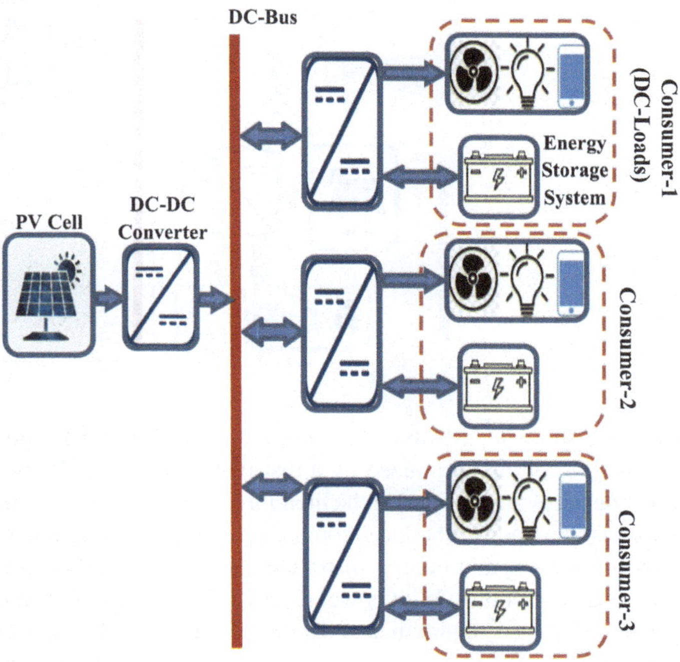

Fig. 4 Typical architecture of CGDSA

household implements. There is distribution losses is associated with architecture, depends on number of users in the remote areas. From the control and monitoring perspective, CGCSA architecture is reliable. However, power losses in distribution is associated. Hence, proving electricity access to communal load are not feasible in CGCSA architecture [16].

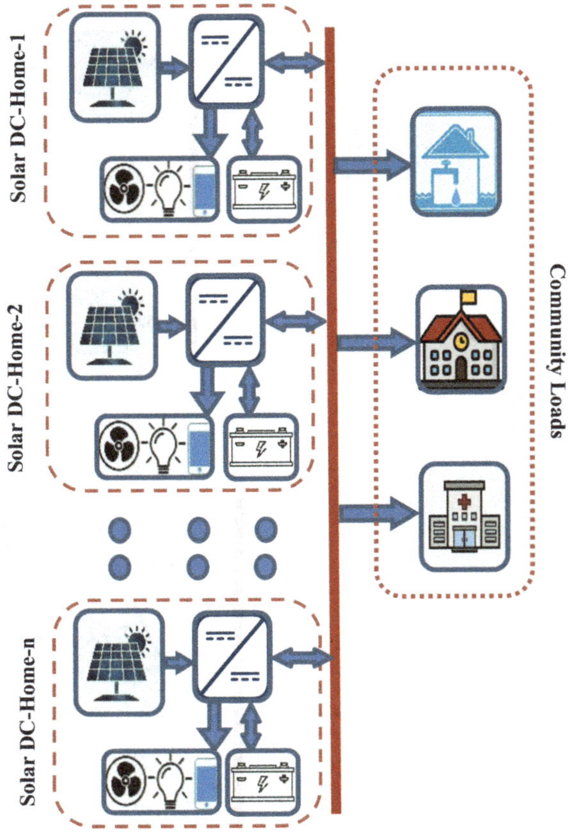

Fig. 5 Typical architecture of DGDSA

Figures 4 and 5 show the distributed architecture or DcMG. The limitation of a centralized structure can be minimized by a distributed structure. There are many distributed architectures are reported in the literature, still work going on modularity in these structures to empower the local communities towards sustainable development. The power losses are minimum in distributed architecture and can deliver more power to individual households. With power sharing and aggregation features, it can tackle the community load requirements. The ESS at every individual users node will led to reduction of distribution losses and available power can be intelligently managed.

The topology of DcMG is depicted on Fig. 6. The SPV is interfaced via a buck converter and battery storage through a bidirectional DC-DC converter. The stability and reliability are significantly improved through grid interaction in DC-bus via the AC-DC converter. The power semiconductors switches of the converter are controlled by a real-time controller. The DC-loads are connected in a common DC-bus of 48 V chosen for safety and have good percentage loading for low voltage applications. The comparison between different EMSs are described in Table 1.

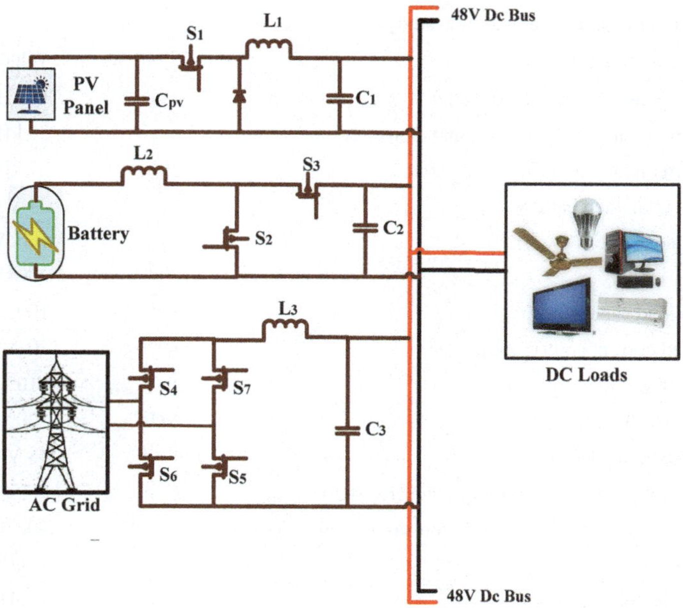

Fig. 6 Topology of DcMG

Table 1 Comparison of various EMS

Attribute	Centralized EMS	Decentralized EMS	Distributed EMS
Controllers	Local controllers (LC) are depends on central controller (CC) decision	LCs are independent to take decision on the basis of terminal quantities	Interaction between LCs to come at control decision
Reliability	Reliability is degraded due to centralized structure	Reliability is sustained	Reliability is sustained
System Failure	Single point of failure can obstruct the whole MG	No dependency on the central controller	No dependency on the central controller
Computational Burden	High	Least	Less than centralized EMS
Plug and play functionality	Poor	Easy	Can be achieved with communication requirements

4 Simulation Setup

Simulation of basic dc system with a dc bus voltage of 48 V DC is made with MATLAB Simulink. The voltage range will be enough to drive most of the DC loads

Table 2 Parameters of simulated case study

Description of the parameters	Symbol	Value
Switching frequency of buck converter	F_{sw}	10 kHz
Switching frequency of Bi-Directional converter	F_{sw}	25 kHz
Switching frequency of AC/DC converter	F_{sw}	25 kHz
Rated voltage of each battery	V^b	24 V
Battery capacity	C_i	100 Ah
PV capacity	P^{pv}	200–500 W
Household load capacity	P^{HL}	100–200 W
Max allowed current contribution	I_i^{max}	20 A
Threshold for community sharing	SOC_{min}	10%
Max utilization Threshold	SOC_{max}	100%
No load voltage for DC bus	V^{NL}	48 V
PI Controller parameters of Battery control system	k_p, k_i	0.25, 20
PI Controller parameters of AC/DC converter control system	k_p, k_i	40, 2000
DC bus Capacitance	C_{bus}	1 mF
Inductor value for buck converter	L	1 mH
The Capacitance of buck converter	C	1000 mF
Inductor value for Bi-directional converter	L	0.576 mH
Inductor value of AC/DC converter	L	0.5 H

like lights, fans, TVs, and some power sockets can also connected. The proposed design MATLAB Simulink model is shown in Fig. 6.

In this model, we are going to use a solar system to generate required dc voltage, and a battery which is sort of like backup when solar is not available and ac grid which is a final resort i.e. when both solar and battery are not available, then ac will drive the system through ac to dc converter. The simulation parameters considered are shown in Table 2.

4.1 SPV System

The PV module produces the electrical energy of the electromagnetic energy of photons referred to as irradiance present in solar radiation at ambient temperature, the magnitude of the irradiance level should be 1000 W/m^2 at standard temperature and the temperature should be 25 °C at this magnitude of the solar irradiance and the temperature, the output voltage obtained is 62 V dc and total power is 1705.2 Wp.

Since DC bus voltage is 48 V we need to step down this 62 V to the required 48 V. For this purpose, a buck converter is installed at PV module output as shown in Fig. 7. Since for simulation purposes, the module inputs are kept constant to get maximum

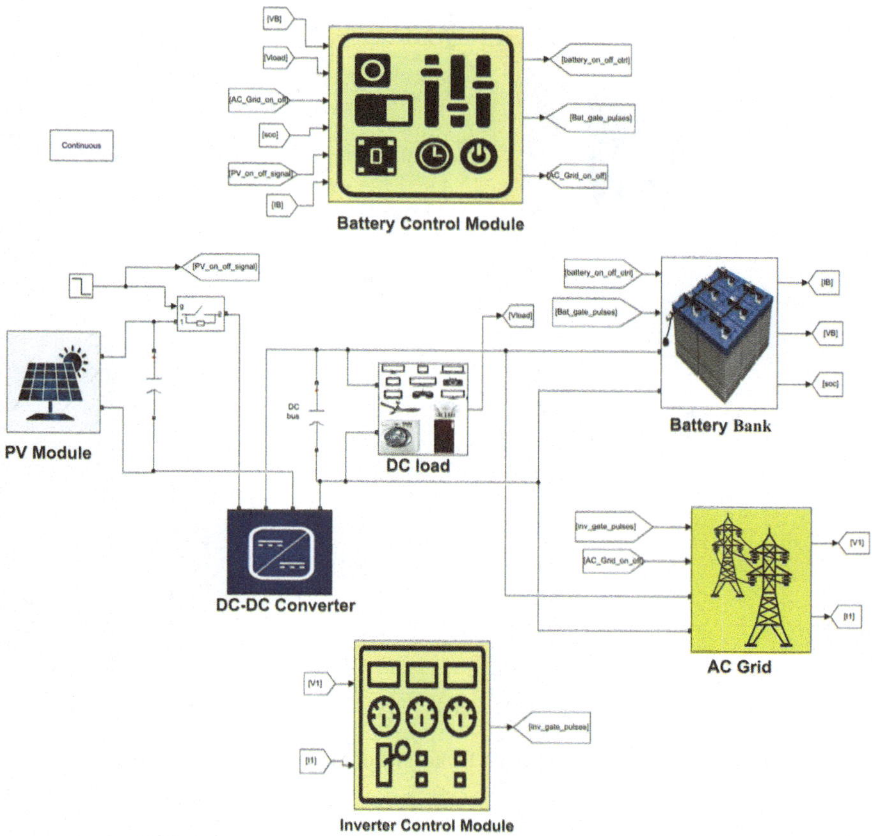

Fig. 7 MATLAB Simulink Model of DcMG

power from the PV module for the full simulation period. So the triggering pulses used for switching of the switch (MOSFET) in the converter is normal DC-DC PWM generator with a 77% duty ratio.

4.2 Energy Storage System

The performance of any battery is determined based on its SOC. This also determines the level amount of power available in the battery. The battery sizing can be done with this by specifying its nominal voltage capacity and the initial level of SOC. since there will be Bi-directional power flow between DC bus and battery both charging and discharging of this battery will occur. So we used a Bi-directional buck-boost converter as shown in Fig. 8. An inductor is placed to smooth the battery current.

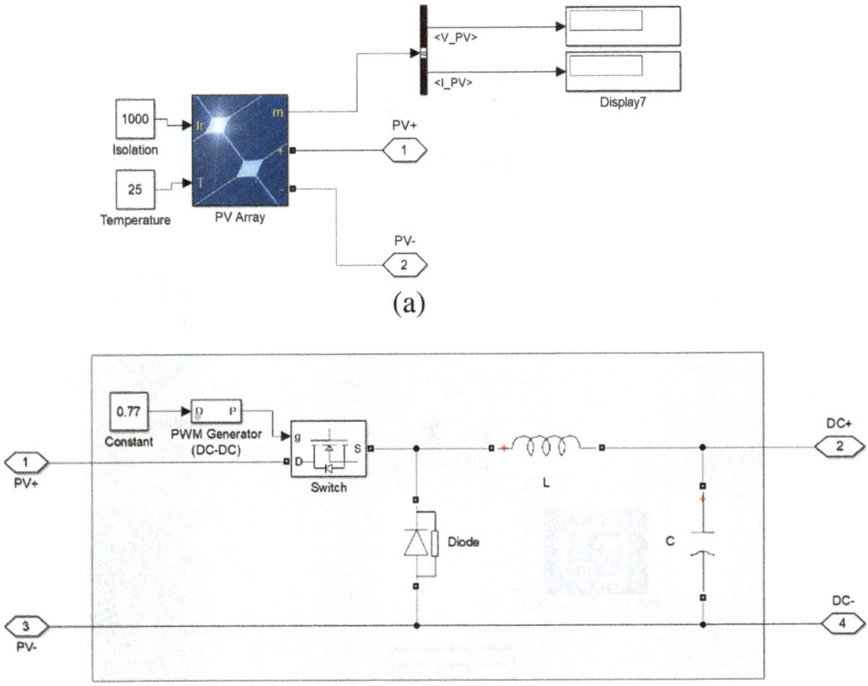

Fig. 8 Simulation diagram of **a** SPV (Solar Photovoltaic) Module **b** Buck Converter

The Closed-loop Constant Current Control Mechanism (CCCM) is used to produce switch gate signals, as shown in Fig. 9, for each mode an individual control loop is built. The battery nominal voltage is 1.2 times the reference for recharge

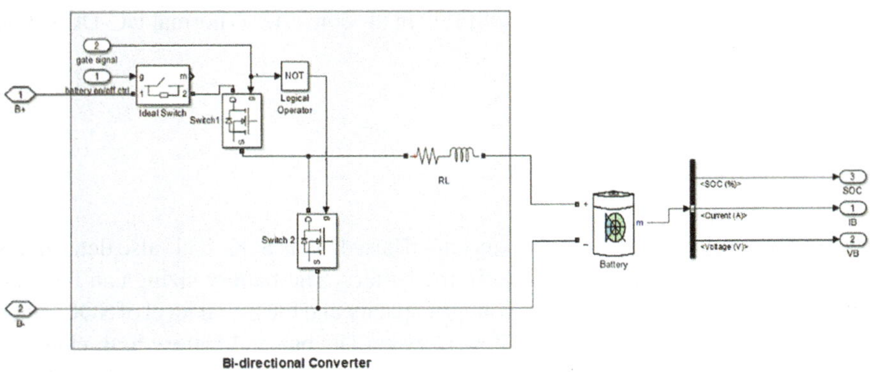

Fig. 9 ESS with interleaved Bi-directional Buck-Boost converter

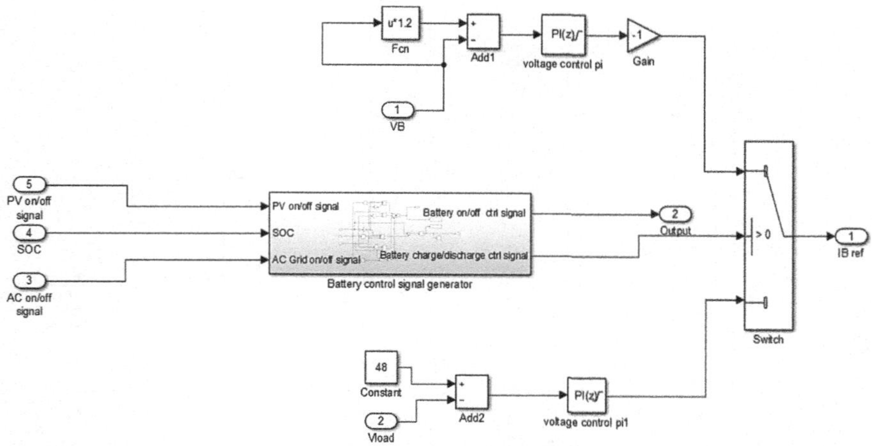

Fig. 10 ESS control system

mode, and the DC buss level voltage for download mode. This is known to improve battery life by increasing the charge and discharge cycles. Therefore, an effective control mechanism is required for battery operations to be charged, discharged and shut off (Fig. 10).

Advanced techniques such as fuzzy logic can be used as a control mechanism for various modes of battery operation, but the control mechanism can slow down the simulation process because in this situation, several conditional probabilities need to be tested and the internal process becomes complex. The used DCMG model is already pretty large and the introduction of fuzzy would complicate it. So here the basic mathematical model-based control mechanism is used to validate various modes of battery operation.

The complete modes of operation and control scheme can be understood by the flow chart shown in Fig. 11. The flow chart depicts the different level of SOC of battery and working of DcMG.

4.3 AC-DC Converter Module

The major problem with SPV is that it's is not available throughout the day for the whole year. During rainy and winter seasons when the sun is not at its peak, the available net energy generation will be very less to meet load demands for the whole day. Also, there are issues like partial shading, panel voltage mismatching conditions, etc. which further affects the energy generation by SPV. The net energy storage capacity hold by the battery is also low due to economic constraints, as an increase in battery size will increase the cost of the overall system.

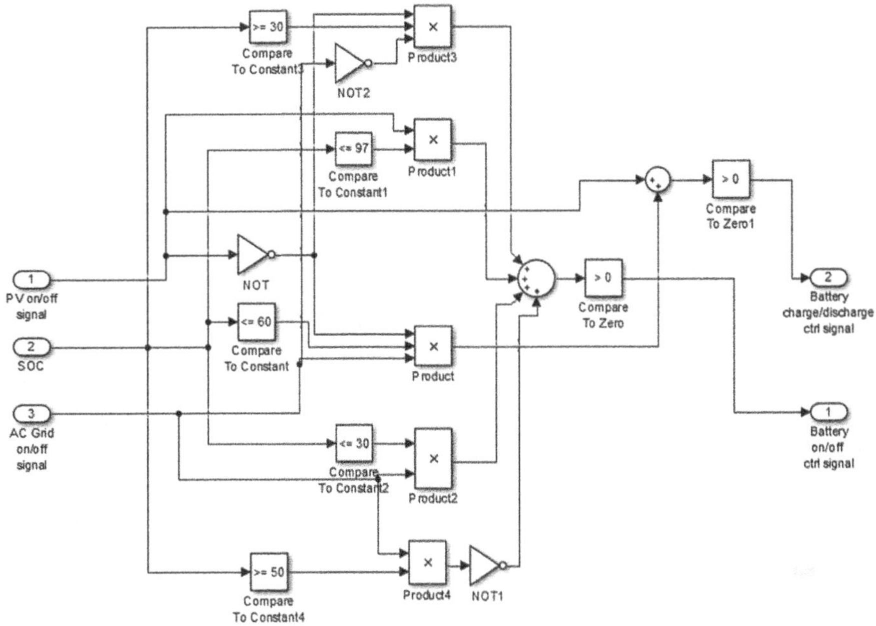

Fig. 11 Block diagram of implemented control mechanism

In this simulation model, an AC utility grid is connected to ensure power availability which comes into play when the solar is off and the battery SOC level is less than 60%.

The Fig. 12 shows the MATLAB/Simulink diagram of an AC utility grid with a rectifier circuit, here Power flow is unidirectional i.e. from ac utility to dc bus. A filter is used to eliminate the harmonics at the output of the ac/dc converter. The closed-loop control mechanism shown in Fig. 13 is used to generate gate pulses for switches connected in a parallel configuration in two fully controlled legs. As the required dc bus voltage is 48Vdc we choose the reference of 48. Here sinusoidal PWM (Pulse Width Modulation) technique is used where the Modulation Index of PWM pulse is used as a control element (Fig. 14).

4.4 DC Load

The major issue that drives the installation of the DcMG in the homes is the big gap between the power consumption among the DC and AC appliances. Nowadays this gap in power is being observed by a lot of users and they are gradually replacing their AC devices with DC devices. Any DC appliance can be typically represented using a resistor. So, a resistive load can also being used as shown in Fig. 15.

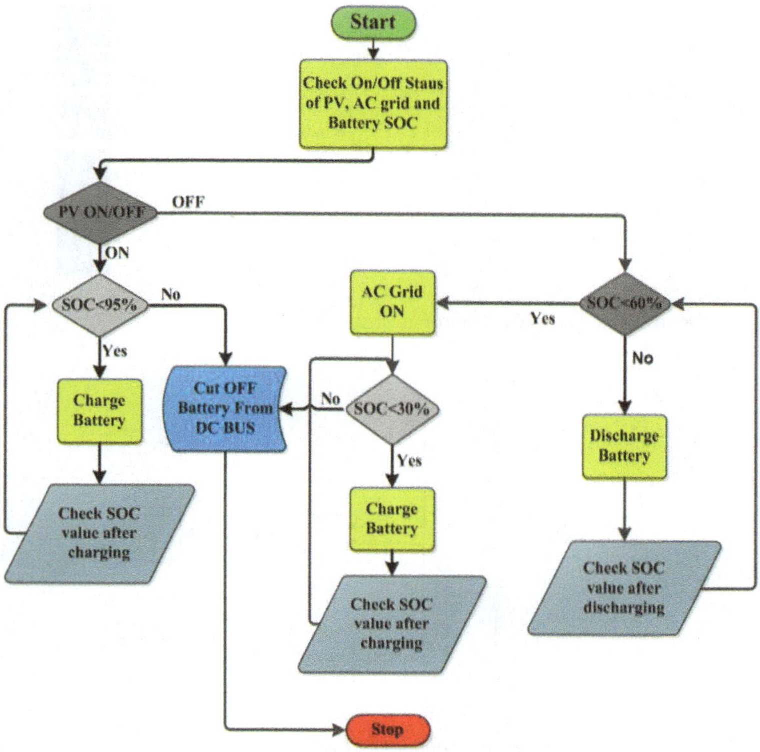

Fig. 12 Flow Chart of control scheme

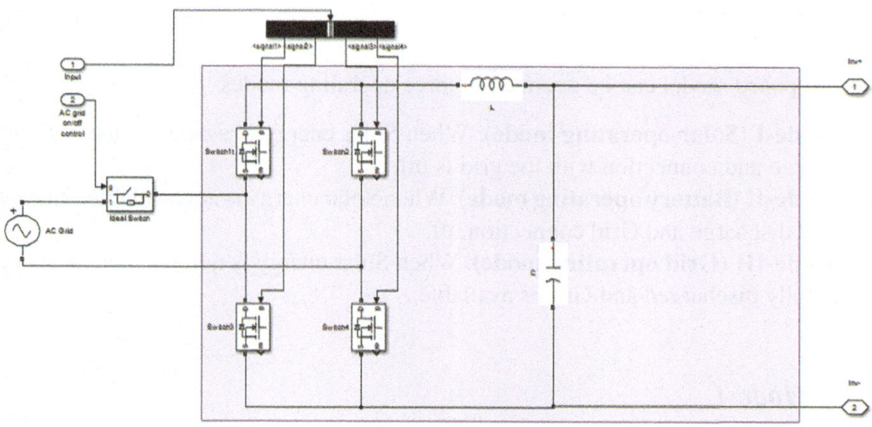

Fig. 13 Rectifier Module with AC grid on/off control switch

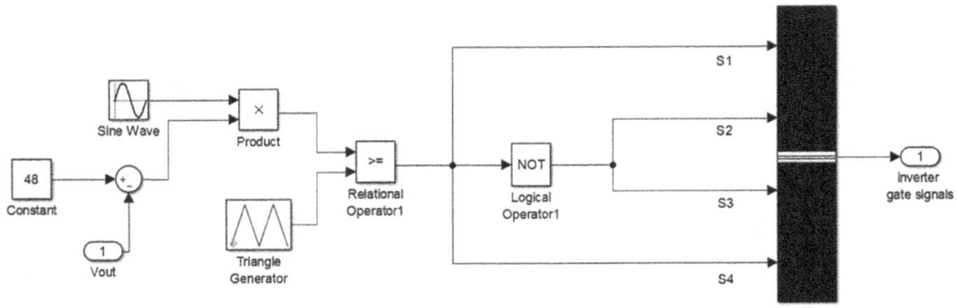

Fig. 14 AC-DC Rectifier gate pulse control mechanism

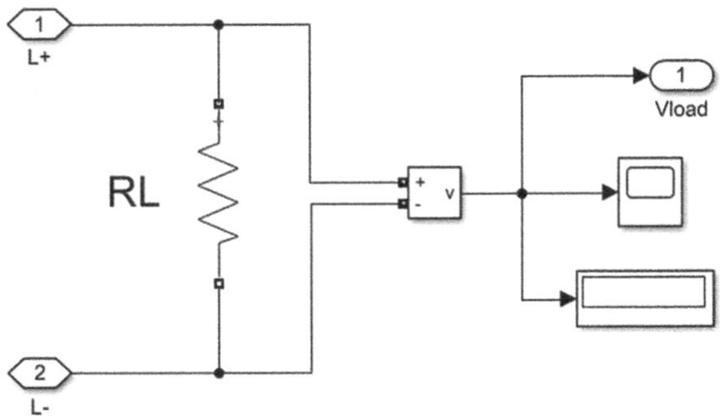

Fig. 15 Equivalent circuit simulation model of DC-load

5 Results and Discussion

The proposed model can be operated in three operating modes

1. **Mode-I (Solar operating mode)**: When Solar energy is available, battery will charge and connection with the grid is off.
2. **Mode-II (Battery operating mode)**: When Solar energy is not available, Battery will discharge and Grid connection off.
3. **Mode-III (Grid operating mode)**: When Solar energy is not available, Battery is fully discharged and Grid is available.

5.1 Mode-I

As explained in the battery control scheme and AC grid control scheme if solar energy is available, the PV panel will supply, then the AC grid will be disconnected and the battery may charge or cut off (which depends on the SOC level of the battery) from the DC bus. The DC bus voltage and SOC level of the battery, in this case, are shown in Fig. 16.

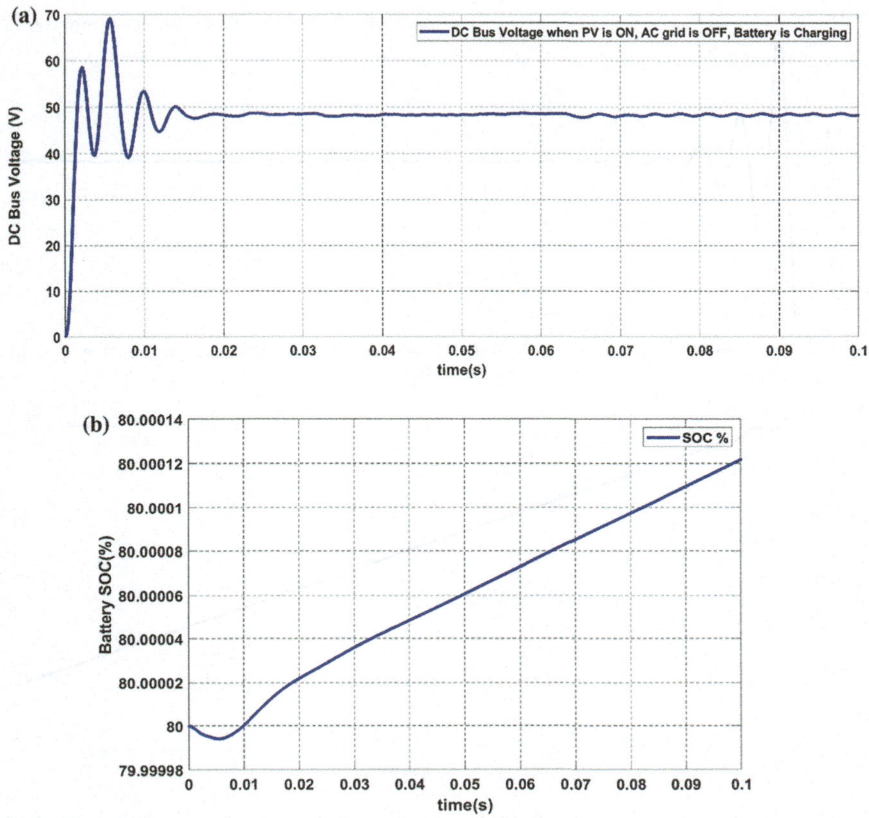

Fig. 16 **a** DC bus voltage at PV is ON, AC Grid is OFF, and Battery is charging **b** SOC of battery at charging mode

5.2 *Mode-II*

Here two subcases arises based on the SOC level i.e. ac grid can be on/off. Depending on the SOC levels and ac grid status the battery may charge, discharge, or can be cut off from the system. Some of these cases are shown in Fig. 17.

5.3 *Mode-III*

When ac grid is available, the battery will charge or be in the cut-off state based on its SOC levels, simulation results of DC bus voltage as shown in Fig. 18.

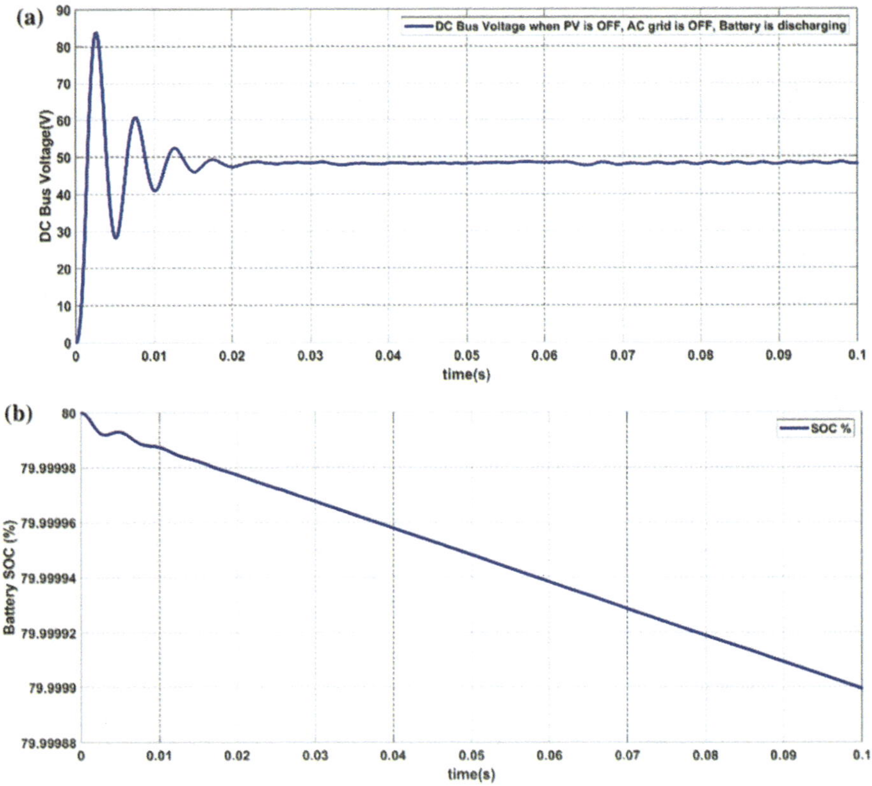

Fig. 17 **a** DC bus voltage at PV is ON, AC Grid is OFF, and Battery is charging **b** SOC of battery at charging mode

6 Recent Trends in DcMG Research

The concept of MG was introduced in [24] which involves dividing the distribution system into several regions. There are extensive research works that are going in the field of structure, control, and integration of DcMG, various EMS. The following are the recent evolving trends in DcMG research.

6.1 Minimization of Effect of PV and Load Variation

The DcMG need to incorporate the PV and load variation for voltage regulation of DC-bus. The control of semiconductor switches in converter must include those variation for optimum energy utilization and ensuring the power continuity at load. It is one of the key challenge to incorporate both variation to stabilize DC-bus voltage.

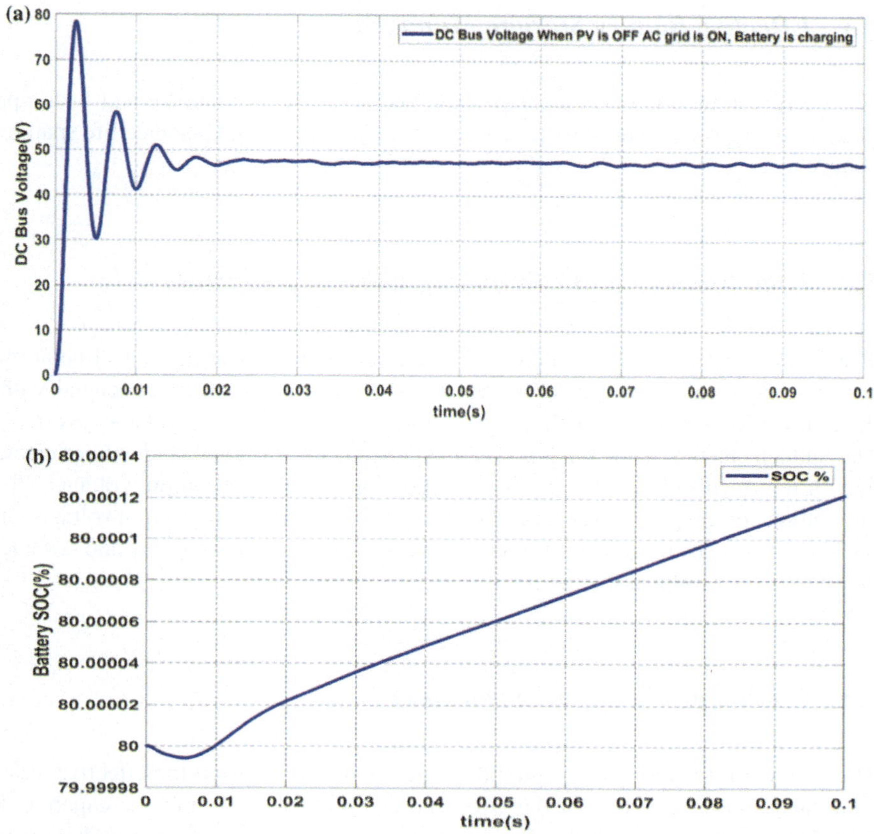

Fig. 18 **a** DC bus voltage at PV is OFF, AC Grid is ON, and Battery is charging **b** SOC of battery at charging mode

6.2 Reduction of Ripple Current

The minimum ripples in DC-bus current ensure the better performance of the load. The losses can also be minimized with minimal ripples in output current. Thus it is an important concern of research to minimize ripples and increase the efficacy of Dc-MG via proper selection of DC-DC converter for PV and storage system, and its control schemes.

6.3 Effective Battery Storage Management

The energy storage is plays an important role in Dc-MG reliability and back-up, thus effective power management required to reduce stress and enhance life span of battery.

6.4 Load Sharing and Circulating Current Reduction

The DcMG control is associated with the issue of load sharing and circulating between the parallel-connected converters. The DcMG control is associated with the issue of load sharing and circulating between the parallel-connected converters. The Sudden changes in the input power, load, parametric variation, and voltage error, and current feedback are the reasons for variations in constant output voltage. The circulating current problem occurs when there is a difference in the output voltages of the converters. The researchers could explore these issues to have stable and reliable DcMG for the rural mass.

6.5 Challenges in DcMG Protection

The DcMG protection scheme should be well planned to ensure the effective, efficient, and secure operation of the grid. The DcMG protection must be capable of reliably detecting and isolating the fault to achieve minimal disruption and outage. Integrating DES into the grid raises chances of system instability. DcMG protection is topologically dependent and affects the key grid protection mechanism. Pole to pole is the potential fault in the DcMG, and pole to ground faults is. Due to non-zero crossing in the DC scheme, the design of DcMG is difficult, there is a huge research opportunity in these areas of DcMG which requires further exploration.

The well-designed control strategy, integration, and intelligent power management between multiple nano-grids under DcMG. The smart power-sharing among nano-grids can enable the community load to access the electricity that provides reliable power solutions to remote areas. In the near future, this concept can be combined with several grids that provide resilient, reliable, and flexible power to the consumers.

7 Conclusion

This chapter reports an overview of the architectural and control aspects of DcMG in the context of remote electrification. There has been a considerable volume of research on various aspects of DcMG. The system is simulated and results are discussed. A comprehensive analysis and comparison of different modes of operation have been presented in the DcMG. The DcMG is gaining interest among the researchers from various perspectives like reliability, minimization of power electronics converters and loss associated with it, and cost-effectiveness. There is a lot of potential scope of research in DcMG for remote electrification in the Indian scenario. The DcMG will play an important role in future energy demand and consider as a future grid.

References

1. F. Blaabjerg, Z. Chen, S.B. Kjaer, Power electronics as efficient interface in dispersed power generation systems. IEEE Trans. Power Electron. **19**(5), 1184–1194 (2004)
2. M. Nasir, H.A. Khan, N.A. Zaffar, J.C. Vasquez, J.M. Guerrero, Scalable solar DC micrigrids: on the path to revolutionizing the electrification architecture of developing communities. IEEE Electron. Mag. **6**, 63–72 (2018)
3. A.D. Riana, F. Hunsnayain, E.A. Pramana, H. Song, P.Y.D. Setyo, A. Zulfia, C. Hudaya, Implementation of talis and dc house system for rural areas in indonesia, in *MATEC Web of Conferences*, vol. 218, 01006 (2018)
4. A. Newcombe, E. Ackom, Sustainable solar home systems model: applying lessons from Bangladesh to Myanmar's rural poor. Energy Sustain. Develop. **38**, 21–33 (2017)
5. M. Fowlie, Y. Khaitan, C. Wolfram, D. Wolfson, Solar microgrids and remote energy access: how weak incentives can undermine smart technology (2018)
6. C. Azimoh, P. Klintenberg, F. Wallin, B. Karlsson, Illuminated but not electrified: an assessment of the impact of solar home system on rural households in South Africa. Appl. Energy **155**, 354–364 (2015)
7. H. Khan, H. Ahmad, M. Nasir, M. Nadeem, N. Zaffar, Decentralised electric power delivery for rural electrification in Pakistan. Energy Policy **120**, 312–323 (2018)
8. M. Nasir, M. Anees, H. Khan, I. Khan, Y. Xu, J. Guerrero, Integration and decentralized control of standalone solar home systems for off-grid community applications. IEEE Trans. Ind. Appl. **55**, 7240–7250 (2019)
9. B. Sovacool, Success and failure in the political economy of solar electrification: lessons from world bank solar home system (SHS) projects in Sri Lanka and Indonesia. Energy Policy **123**, 482–493 (2018)
10. F. Ahmad, M. Alam, Economic and ecological aspects for microgrids deployment in India. Sustain. Cities Soc. **37**, 407–419 (2018)
11. A. Kumar, R. Ferdous, A. Luque-Ayala, C. McEwan, M. Power, B. Turner, H. Bulkeley, Solar energy for all? Understanding the successes and shortfalls through a critical comparative assessment of Bangladesh, Brazil, India, Mozambique, Sri Lanka and South Africa. Energy Res. Soc. Sci. **48**, 166–176 (2019)
12. M. Nasir, H. Khan, N. Zaffar, J. Vasquez, J. Guerrero, Scalable solar dc micrigrids: on the path to revolutionizing the electrification architecture of developing communities. IEEE Electr. Mag. **6**, 63–72 (2018)
13. J. Guerrero, J. Vasquez, J. Matas, L. de Vicuna, M. Castilla, Hierarchical control of droop-controlled AC and DC microgrids—a general approach toward standardization. IEEE Trans. Ind. Electron. **58**, 158–172 (2011)
14. D. Palit, S. Malhotra, Energizing rural India using micro grids: the case of solar DC micro-grids in Uttar Pradesh State, India in Micro Perspectives for Decentralized Energy Supply, in *Proceedings of the International Conference*, vol. 1, pp. 56–60 (2015)
15. M. Millinger, T. Mårlind, E. Ahlgren, Evaluation of Indian rural solar electrification: a case study in Chhattisgarh. Energy Sustain. Develop. **16**, 486–492 (2012)

16. S. Sahoo, A. Sinha, N. Kishore, Control techniques in AC, DC, and hybrid AC–DC microgrid: a review. IEEE J. Emerg. Sel. Top. Power Electron. **6**, 738–759 (2018)
17. A. Ipakchi, F. Albuyeh, Grid of the future. IEEE Power Energ. Mag. **7**, 52–62 (2009)
18. H. Farhangi, The path of the smart grid. IEEE Power Energ. Mag. **8**, 18–28 (2010)
19. T. Dragicevic, X. Lu, J. Vasquez, J. Guerrero, DC microgrids—part ii: a review of power architectures, applications, and standardization issues. IEEE Trans. Power Electron. **31**, 3528–3549 (2016)
20. P. Fairley, DC versus AC: the second war of currents has already begun [in my view]. IEEE Power Energ. Mag. **10**, 104–103 (2012)
21. E. Rodriguez-Diaz, J. Vasquez, J. Guerrero, Intelligent DC homes in future sustainable energy systems: when efficiency and intelligence work together. IEEE Consum. Electron. Mag. **5**, 74–80 (2016)
22. A. Jhunjhunwala, A. Lolla, P. Kaur, Solar-dc microgrid for Indian homes: a transforming power scenario. IEEE Electri. Mag. **4**, 10–19 (2016)
23. U. Manandhar, B. Wang, X. Zhang, G. Beng, Y. Liu, A. Ukil, Joint control of three-level DC–DC converter interfaced hybrid energy storage system in DC microgrids. IEEE Trans. Energy Convers. **34**, 2248–2257 (2019)
24. E. Ng, R. El-Shatshat, Multi-microgrid control systems (MMCS), in *IEEE PES General Meeting* (2010)

The manufacturer's authorised representative in the EU is Springer Nature Customer Service Centre GmbH, Europaplatz 3, 69115 Heidelberg, Germany. If you have any concerns regarding our products, please contact ProductSafety@springernature.com

Printed and bound by CPI Group (UK) Ltd, Croydon, CR0 4YY
30/09/2026
02225291-0007